FREE Test Taking Tips DVD Offer

To help us better serve you, we have developed a Test Taking Tips DVD that we would like to give you for FREE. **This DVD covers world-class test taking tips that you can use to be even more successful when you are taking your test.**

All that we ask is that you email us your feedback about your study guide. Please let us know what you thought about it – whether that is good, bad or indifferent.

To get your **FREE Test Taking Tips DVD**, email freedvd@studyguideteam.com with "FREE DVD" in the subject line and the following information in the body of the email:

a. The title of your study guide.

b. Your product rating on a scale of 1-5, with 5 being the highest rating.

c. Your feedback about the study guide. What did you think of it?

d. Your full name and shipping address to send your free DVD.

If you have any questions or concerns, please don't hesitate to contact us at freedvd@studyguideteam.com.

Thanks again!

Rhea

FSA Math Grade 7 Workbook

FSA Math 7th Grade for Florida Standards Assessment [7th Grade Math Workbook]

Test Prep Books

Table of Contents

Quick Overview

As you draw closer to taking your exam, effective preparation becomes more and more important. Thankfully, you have this study guide to help you get ready. Use this guide to help keep your studying on track and refer to it often.

This study guide contains several key sections that will help you be successful on your exam. The guide contains tips for what you should do the night before and the day of the test. Also included are test-taking tips. Knowing the right information is not always enough. Many well-prepared test takers struggle with exams. These tips will help equip you to accurately read, assess, and answer test questions.

A large part of the guide is devoted to showing you what content to expect on the exam and to helping you better understand that content. In this guide are practice test questions so that you can see how well you have grasped the content. Then, answer explanations are provided so that you can understand why you missed certain questions.

Don't try to cram the night before you take your exam. This is not a wise strategy for a few reasons. First, your retention of the information will be low. Your time would be better used by reviewing information you already know rather than trying to learn a lot of new information. Second, you will likely become stressed as you try to gain a large amount of knowledge in a short amount of time. Third, you will be depriving yourself of sleep. So be sure to go to bed at a reasonable time the night before. Being well-rested helps you focus and remain calm.

Be sure to eat a substantial breakfast the morning of the exam. If you are taking the exam in the afternoon, be sure to have a good lunch as well. Being hungry is distracting and can make it difficult to focus. You have hopefully spent lots of time preparing for the exam. Don't let an empty stomach get in the way of success!

When travelling to the testing center, leave earlier than needed. That way, you have a buffer in case you experience any delays. This will help you remain calm and will keep you from missing your appointment time at the testing center.

Be sure to pace yourself during the exam. Don't try to rush through the exam. There is no need to risk performing poorly on the exam just so you can leave the testing center early. Allow yourself to use all of the allotted time if needed.

Remain positive while taking the exam even if you feel like you are performing poorly. Thinking about the content you should have mastered will not help you perform better on the exam.

Once the exam is complete, take some time to relax. Even if you feel that you need to take the exam again, you will be well served by some down time before you begin studying again. It's often easier to convince yourself to study if you know that it will come with a reward!

Test-Taking Strategies

1. Predicting the Answer

When you feel confident in your preparation for a multiple-choice test, try predicting the answer before reading the answer choices. This is especially useful on questions that test objective factual knowledge. By predicting the answer before reading the available choices, you eliminate the possibility that you will be distracted or led astray by an incorrect answer choice. You will feel more confident in your selection if you read the question, predict the answer, and then find your prediction among the answer choices. After using this strategy, be sure to still read all of the answer choices carefully and completely. If you feel unprepared, you should not attempt to predict the answers. This would be a waste of time and an opportunity for your mind to wander in the wrong direction.

2. Reading the Whole Question

Too often, test takers scan a multiple-choice question, recognize a few familiar words, and immediately jump to the answer choices. Test authors are aware of this common impatience, and they will sometimes prey upon it. For instance, a test author might subtly turn the question into a negative, or he or she might redirect the focus of the question right at the end. The only way to avoid falling into these traps is to read the entirety of the question carefully before reading the answer choices.

3. Looking for Wrong Answers

Long and complicated multiple-choice questions can be intimidating. One way to simplify a difficult multiple-choice question is to eliminate all of the answer choices that are clearly wrong. In most sets of answers, there will be at least one selection that can be dismissed right away. If the test is administered on paper, the test taker could draw a line through it to indicate that it may be ignored; otherwise, the test taker will have to perform this operation mentally or on scratch paper. In either case, once the obviously incorrect answers have been eliminated, the remaining choices may be considered. Sometimes identifying the clearly wrong answers will give the test taker some information about the correct answer. For instance, if one of the remaining answer choices is a direct opposite of one of the eliminated answer choices, it may well be the correct answer. The opposite of obviously wrong is obviously right! Of course, this is not always the case. Some answers are obviously incorrect simply because they are irrelevant to the question being asked. Still, identifying and eliminating some incorrect answer choices is a good way to simplify a multiple-choice question.

4. Don't Overanalyze

Anxious test takers often overanalyze questions. When you are nervous, your brain will often run wild, causing you to make associations and discover clues that don't actually exist. If you feel that this may be a problem for you, do whatever you can to slow down during the test. Try taking a deep breath or counting to ten. As you read and consider the question, restrict yourself to the particular words used by the author. Avoid thought tangents about what the author *really* meant, or what he or she was *trying* to say. The only things that matter on a multiple-choice test are the words that are actually in the question. You must avoid reading too much into a multiple-choice question, or supposing that the writer meant something other than what he or she wrote.

5. No Need for Panic

It is wise to learn as many strategies as possible before taking a multiple-choice test, but it is likely that you will come across a few questions for which you simply don't know the answer. In this situation, avoid panicking. Because most multiple-choice tests include dozens of questions, the relative value of a single wrong answer is small. As much as possible, you should compartmentalize each question on a multiple-choice test. In other words, you should not allow your feelings about one question to affect your success on the others. When you find a question that you either don't understand or don't know how to answer, just take a deep breath and do your best. Read the entire question slowly and carefully. Try rephrasing the question a couple of different ways. Then, read all of the answer choices carefully. After eliminating obviously wrong answers, make a selection and move on to the next question.

6. Confusing Answer Choices

When working on a difficult multiple-choice question, there may be a tendency to focus on the answer choices that are the easiest to understand. Many people, whether consciously or not, gravitate to the answer choices that require the least concentration, knowledge, and memory. This is a mistake. When you come across an answer choice that is confusing, you should give it extra attention. A question might be confusing because you do not know the subject matter to which it refers. If this is the case, don't eliminate the answer before you have affirmatively settled on another. When you come across an answer choice of this type, set it aside as you look at the remaining choices. If you can confidently assert that one of the other choices is correct, you can leave the confusing answer aside. Otherwise, you will need to take a moment to try to better understand the confusing answer choice. Rephrasing is one way to tease out the sense of a confusing answer choice.

7. Your First Instinct

Many people struggle with multiple-choice tests because they overthink the questions. If you have studied sufficiently for the test, you should be prepared to trust your first instinct once you have carefully and completely read the question and all of the answer choices. There is a great deal of research suggesting that the mind can come to the correct conclusion very quickly once it has obtained all of the relevant information. At times, it may seem to you as if your intuition is working faster even than your reasoning mind. This may in fact be true. The knowledge you obtain while studying may be retrieved from your subconscious before you have a chance to work out the associations that support it. Verify your instinct by working out the reasons that it should be trusted.

8. Key Words

Many test takers struggle with multiple-choice questions because they have poor reading comprehension skills. Quickly reading and understanding a multiple-choice question requires a mixture of skill and experience. To help with this, try jotting down a few key words and phrases on a piece of scrap paper. Doing this concentrates the process of reading and forces the mind to weigh the relative importance of the question's parts. In selecting words and phrases to write down, the test taker thinks about the question more deeply and carefully. This is especially true for multiple-choice questions that are preceded by a long prompt.

9. Subtle Negatives

One of the oldest tricks in the multiple-choice test writer's book is to subtly reverse the meaning of a question with a word like *not* or *except*. If you are not paying attention to each word in the question, you can easily be led astray by this trick. For instance, a common question format is, "Which of the following is...?" Obviously, if the question instead is, "Which of the following is not...?," then the answer will be quite different. Even worse, the test makers are aware of the potential for this mistake and will include one answer choice that would be correct if the question were not negated or reversed. A test taker who misses the reversal will find what he or she believes to be a correct answer and will be so confident that he or she will fail to reread the question and discover the original error. The only way to avoid this is to practice a wide variety of multiple-choice questions and to pay close attention to each and every word.

10. Reading Every Answer Choice

It may seem obvious, but you should always read every one of the answer choices! Too many test takers fall into the habit of scanning the question and assuming that they understand the question because they recognize a few key words. From there, they pick the first answer choice that answers the question they believe they have read. Test takers who read all of the answer choices might discover that one of the latter answer choices is actually *more* correct. Moreover, reading all of the answer choices can remind you of facts related to the question that can help you arrive at the correct answer. Sometimes, a misstatement or incorrect detail in one of the latter answer choices will trigger your memory of the subject and will enable you to find the right answer. Failing to read all of the answer choices is like not reading all of the items on a restaurant menu: you might miss out on the perfect choice.

11. Spot the Hedges

One of the keys to success on multiple-choice tests is paying close attention to every word. This is never truer than with words like almost, most, some, and sometimes. These words are called "hedges" because they indicate that a statement is not totally true or not true in every place and time. An absolute statement will contain no hedges, but in many subjects, the answers are not always straightforward or absolute. There are always exceptions to the rules in these subjects. For this reason, you should favor those multiple-choice questions that contain hedging language. The presence of qualifying words indicates that the author is taking special care with his or her words, which is certainly important when composing the right answer. After all, there are many ways to be wrong, but there is only one way to be right! For this reason, it is wise to avoid answers that are absolute when taking a multiple-choice test. An absolute answer is one that says things are either all one way or all another. They often include words like *every*, *always*, *best*, and *never*. If you are taking a multiple-choice test in a subject that doesn't lend itself to absolute answers, be on your guard if you see any of these words.

12. Long Answers

In many subject areas, the answers are not simple. As already mentioned, the right answer often requires hedges. Another common feature of the answers to a complex or subjective question are qualifying clauses, which are groups of words that subtly modify the meaning of the sentence. If the question or answer choice describes a rule to which there are exceptions or the subject matter is complicated, ambiguous, or confusing, the correct answer will require many words in order to be expressed clearly and accurately. In essence, you should not be deterred by answer choices that seem excessively long. Oftentimes, the author of the text will not be able to write the correct answer without offering some qualifications and modifications. Your job is to read the answer choices thoroughly and completely and to select the one that most accurately and precisely answers the question.

13. Restating to Understand

Sometimes, a question on a multiple-choice test is difficult not because of what it asks but because of how it is written. If this is the case, restate the question or answer choice in different words. This process serves a couple of important purposes. First, it forces you to concentrate on the core of the question. In order to rephrase the question accurately, you have to understand it well. Rephrasing the question will concentrate your mind on the key words and ideas. Second, it will present the information to your mind in a fresh way. This process may trigger your memory and render some useful scrap of information picked up while studying.

14. True Statements

Sometimes an answer choice will be true in itself, but it does not answer the question. This is one of the main reasons why it is essential to read the question carefully and completely before proceeding to the answer choices. Too often, test takers skip ahead to the answer choices and look for true statements. Having found one of these, they are content to select it without reference to the question above. Obviously, this provides an easy way for test makers to play tricks. The savvy test taker will always read the entire question before turning to the answer choices. Then, having settled on a correct answer choice, he or she will refer to the original question and ensure that the selected answer is relevant. The mistake of choosing a correct-but-irrelevant answer choice is especially common on questions related to specific pieces of objective knowledge. A prepared test taker will have a wealth of factual knowledge at his or her disposal, and should not be careless in its application.

15. No Patterns

One of the more dangerous ideas that circulates about multiple-choice tests is that the correct answers tend to fall into patterns. These erroneous ideas range from a belief that B and C are the most common right answers, to the idea that an unprepared test-taker should answer "A-B-A-C-A-D-A-B-A." It cannot be emphasized enough that pattern-seeking of this type is exactly the WRONG way to approach a multiple-choice test. To begin with, it is highly unlikely that the test maker will plot the correct answers according to some predetermined pattern. The questions are scrambled and delivered in a random order. Furthermore, even if the test maker was following a pattern in the assignation of correct answers, there is no reason why the test taker would know which pattern he or she was using. Any attempt to discern a pattern in the answer choices is a waste of time and a distraction from the real work of taking the test. A test taker would be much better served by extra preparation before the test than by reliance on a pattern in the answers.

FREE DVD OFFER

Don't forget that doing well on your exam includes both understanding the test content and understanding how to use what you know to do well on the test. We offer a completely FREE Test Taking Tips DVD that covers world class test taking tips that you can use to be even more successful when you are taking your test.

All that we ask is that you email us your feedback about your study guide. To get your **FREE Test Taking Tips DVD**, email freedvd@studyguideteam.com with "FREE DVD" in the subject line and the following information in the body of the email:

- The title of your study guide.
- Your product rating on a scale of 1-5, with 5 being the highest rating.
- Your feedback about the study guide. What did you think of it?
- Your full name and shipping address to send your free DVD.

Introduction

Function of the Test

The Florida Standards Assessments (FSA) Grade 7 Mathematics exam is intended to assess seventh graders' mastery and achievement of Florida math standards. It is part of the compendium of FSA exams, which students take from grade 3 through grade 8 in the subject of math and from grade 3 through grade 10 in the subject of ELA. Tests are administered near the end of the academic year, as they are designed to evaluate the knowledge and skills obtained over the student's current academic grade level. Middle school students who are enrolled in either Algebra 1 or Geometry, rather than a standard seventh grade math course, do not take the FSA Grade 7 Mathematics test. Instead, they take the corresponding End-of-Course (EOC) Assessment.

Test Administration

The FSA Grade 7 Mathematics test is typically given to seventh grade students at school during the first two weeks of May on the specific state-determined dates. It is administered over two days, with a 60-minute session the first day and two more 60-minute sessions the second day, for a total of three 60-minute sessions. It is offered in a computer-based format with a variety of question types including Technology Enhanced Items (TEIs). TEIs are more interactive, requiring the student to construct, select, or support their responses.

Accommodations are available for students with documented disabilities or who are English Language Learners if they have an IEP or Section 504 plan. Examples of available accommodations include large-print text booklets and booklets printed in braille.

On-screen scientific calculators are provided for the two sessions on the second testing day. Handheld scientific calculators are allowed for students with approved accommodations taking a paper-based exam.

Test Format

The test is conducted in three 60-minute sessions over two days (one the first day and two the second day) for a total test time of 180 minutes. There are around 50-60 questions on the test. The test items are designed to measure achievement of the Grade 7 math standards set forth by the state of Florida. Questions may be of a variety of formats including multiple choice, multiselect, Graphic Response Item Display (GRID), and an equation editor format where students can create a response using a toolbar with mathematical symbols. The TEI questions might include hot text items, where students can highlight and select or drag and drop certain responses; open response passages, where students need to write an explanation or defense for their answer; matching exercises; tables to fill in; among others.

The following table details the category and percentage of questions encountered on the test as well as the number of standards pertaining to each category:

Category	Percentage of Questions	# of Standards Assessed
Ratio and Proportional Relationships	25%	3
Expressions and Equations	21%	4
Geometry	23%	6
Statistics and Probability	16%	8
The Number System	15%	3

Scoring

Student performance on the Grade 7 Mathematics exam is categorized into five achievement levels. Level 1 achievement, which correlates to a scaled score of 269-315, indicates that a student's mastery of the standards is "inadequate" and that they are highly likely to need a substantial amount of support in the eighth grade. Level 2 achievement, which correlates to a scaled score of 316-329, is deemed "below satisfactory," and indicates that the student is likely to need a substantial amount of support in the eighth grade. Level 3 achievement, which correlates to a scaled score of 330-345, is considered "satisfactory," and indicates that the student may need support in the eighth grade. Achieving Level 4, which correlates to a scaled score of 346-359, is considered "proficient," and shows the student is likely to excel in the eighth grade. Lastly, Level 5 achievement, which correlates to a scaled score of 360-391, is considered "mastery" and indicative of a student who is highly likely to excel in the eighth grade. The passing score is 330, which is the lowest score in the Level 3 score range.

In 2018, 180,509 students took the FSA Grade 7 Mathematics exam. The mean score was 330, and the percentage of students achieving level 3 or above was 54%. The percentage of seventh graders in each achievement level broke down as follows: Level 1: 26%, Level 2: 20%, Level 3: 26%, Level 4: 17%, Level 5: 11%.

Ratios & Proportional Relationships

Analyzing Proportional Relationships and Using Them to Solve Real-World Problems

Computing Unit Rates

Unit rate word problems will ask to calculate the rate or quantity of something in a different value. For example, a problem might say that a car drove a certain number of miles in a certain number of minutes and then ask how many miles per hour the car was traveling. These questions involve solving proportions. Consider the following examples:

1) Alexandra made \$96 during the first 3 hours of her shift as a temporary worker at a law office. She will continue to earn money at this rate until she finishes in 5 more hours. How much does Alexandra make per hour? How much will Alexandra have made at the end of the day?

$\frac{96}{3} = \frac{x}{8}$ $32 = \frac{x}{8}$ $32 \times 8 = x$ $x = 256$

This problem can be solved in two ways. The first is to set up a proportion, as the rate of pay is constant. The second is to determine her hourly rate, multiply the 5 hours by that rate, and then add the \$96.

To set up a proportion, put the money already earned over the hours already worked on one side of an equation. The other side has x over 8 hours (the total hours worked in the day). It looks like this:

$$\frac{96}{3} = \frac{x}{8}$$

Now, cross-multiply to get $768 = 3x$. To get x, divide by 3, which leaves $x = 256$. Alternatively, as x is the numerator of one of the proportions, multiplying by its denominator will reduce the solution by one step. Thus, Alexandra will make \$256 at the end of the day. To calculate her hourly rate, divide the total by 8, giving \$32 per hour.

Alternatively, it is possible to figure out the hourly rate by dividing \$96 by 3 hours to get \$32 per hour. Now her total pay can be figured by multiplying \$32 per hour by 8 hours, which comes out to \$256.

not needed

2) Jonathan is reading a novel. So far, he has read 215 of the 335 total pages. It takes Jonathan 25 minutes to read 10 pages, and the rate is constant. How long does it take Jonathan to read one page? How much longer will it take him to finish the novel? Express the answer in time.

25/10 = 2.5 min per page

To calculate how long it takes Jonathan to read one page, divide the 25 minutes by 10 pages to determine the page per minute rate. Thus, it takes 2.5 minutes to read one page.

120 pg left × 2.5 = 300 = 5 hrs

Jonathan must read 120 more pages to complete the novel. (This is calculated by subtracting the pages already read from the total.) Now, multiply his rate per page by the number of pages. Thus, $120 \cdot 2.5 = 300$. Expressed in time, 300 minutes is equal to 5 hours.

3) At a hotel, $\frac{4}{5}$ of the 120 rooms are booked for Saturday. On Sunday, $\frac{3}{4}$ of the rooms are booked. On which day are more of the rooms booked, and by how many more?

$\frac{4}{5} \times 120 = 96$ rooms

$\frac{3}{4} \times 120 = 90$ rooms

1) saturday

2) 6 rooms more

The first step is to calculate the number of rooms booked for each day. Do this by multiplying the fraction of the rooms booked by the total number of rooms.

Saturday: $\frac{4}{5} \times 120 = \frac{4}{5} \times \frac{120}{1} = \frac{480}{5} = 96$ rooms

Sunday: $\frac{3}{4} \times 120 = \frac{3}{4} \times \frac{120}{1} = \frac{360}{4} = 90$ rooms

Thus, more rooms were booked on Saturday by 6 rooms.

4) In a veterinary hospital, the veterinarian-to-pet ratio is 1:9. The ratio is always constant. If there are 45 pets in the hospital, how many veterinarians are currently in the veterinary hospital?

Set up a proportion to solve for the number of veterinarians: $\frac{1}{9} = \frac{x}{45}$

Cross-multiplying results in $9x = 45$, which works out to 5 veterinarians.

Alternatively, as there are always 9 times as many pets as veterinarians, is it possible to divide the number of pets (45) by 9. This also arrives at the correct answer of 5 veterinarians.

5) At a general practice law firm, 30% of the lawyers work solely on tort cases. If 9 lawyers work solely on tort cases, how many lawyers work at the firm?

First, solve for the total number of lawyers working at the firm, which will be represented here with x. The problem states that 9 lawyers work solely on torts cases, and they make up 30% of the total lawyers at the firm. Thus, 30% multiplied by the total, x, will equal 9. Written as equation, this is:

$$30\% \times x = 9$$

It's easier to deal with the equation after converting the percentage to a decimal, leaving $0.3x = 9$. Thus, $x = \frac{9}{0.3} = 30$ lawyers working at the firm.

6) Xavier was hospitalized with pneumonia. He was originally given 35mg of antibiotics. Later, after his condition continued to worsen, Xavier's dosage was increased to 60mg. What was the percent increase of the antibiotics? Round the percentage to the nearest tenth.

An increase or decrease in percentage can be calculated by dividing the difference in amounts by the original amount and multiplying by 100. Written as an equation, the formula is:

$$\frac{new\ quantity\ - old\ quantity}{old\ quantity} \times 100$$

Here, the question states that the dosage was increased from 35mg to 60mg, so these are plugged into the formula to find the percentage increase.

$$\frac{60\ -\ 35}{35} \times 100 = \frac{25}{35} \times 100 = .7142\ \times 100\ = 71.4\%$$

Representing Proportional Relationships Between Quantities

Much like a scale factor can be written using an equation like $2A = B$, a **relationship** is represented by the equation $Y = kX$. X and Y are **proportional** because as values of X increase, the values of Y also increase. A relationship that is **inversely proportional** can be represented by the equation $Y = \frac{k}{X}$, where the value of Y decreases as the value of x increases and vice versa.

Deciding Whether Two Quantities are in a Proportional Relationship

Two quantities are **proportional** if there exists a constant ratio between them. The constant, k, is known as the constant of proportionality and proportionality can be expressed in equation form as $y = kx$ or $\frac{y}{x} = k$. Basically, if one quantity is proportional to another, one quantity is a constant multiple of the other. For instance, consider the scenario in which you want to double the number of employees in all of the ice cream stands that you own. This would be represented in equation form as $y = 2x$, where x represents current number of employees and y represents future number of employees. Here is a table that shows 4 scenarios. Notice how the y-value is equal to the x-value doubled.

x	y
1	2
2	4
3	6
4	8

The graph of the equation $y = 2x$ can be seen here:

The graph of the equation $y = 2x$

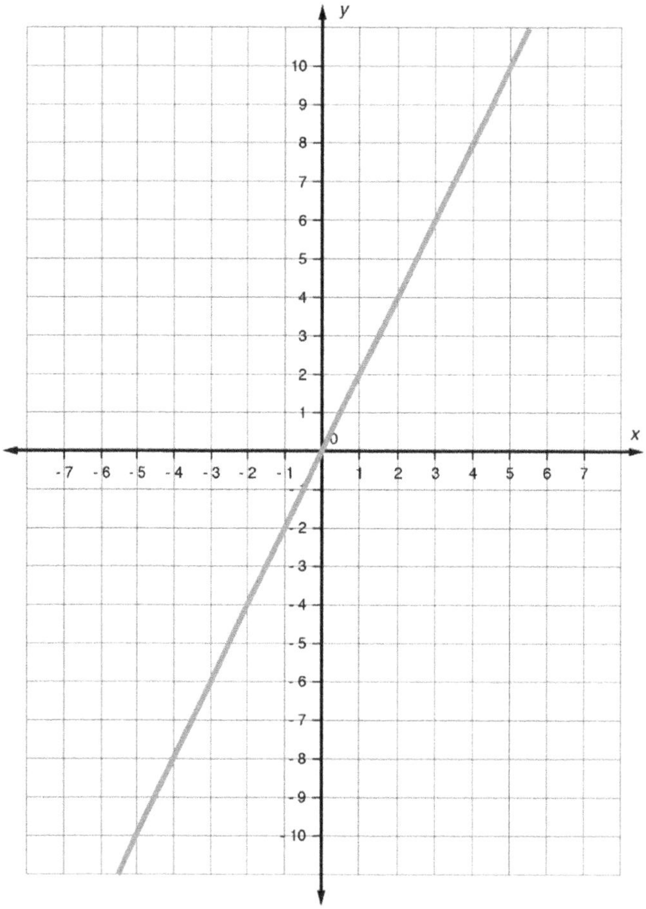

Identifying the Constant of Proportionality in Tables, Graphs, and Equations

Linear relationships may be represented by a table of 2 corresponding values. Certain tables may determine the relationship between the values and predict other corresponding sets. From the table, the constant of proportionality—or the unit rate—can be determined. Proportional relationships will present as a pattern. A simple input-output table can model a pattern that pertains to a specific situation or equation. These can then be utilized in other areas in math, such as graphing.

For example, for every 1 parakeet the pet store sells, it sells 5 goldfish. Using the following equation to model this situation, fill in numbers missing in the input-output table, to show the total number of pets sold by the store.

Equation:

Total number of pets sold (t) = number of parakeets (p) + number of parakeets (p) × 5 goldfish

$$t = p + (p \times 5)$$
$$t = 6p$$

p	t
1	6
2	12
3	18
4	24
5	30

The missing numbers are 18 and 30.

This can also be shown by using an equation. If 3 is put in for p, it would look as follows:

$$t = 6 \times 3$$

$$t = 18$$

If 5 is put in for p, it would look as follows:

$$t = 6 \times 5$$

$$t = 30$$

The completed table would appear as follows:

p	t
1	6
2	12
3	18
4	24
5	30

By looking at the completed table, the numeric patterns between consecutive p and t values (from one row to the next) can be seen. The p-values increase by 1, and the t-values increase by 6.

Let's look at another example of a table showing a linear relationship. Consider the table below, which displays the money in a checking account that charges a monthly fee:

Month	0	1	2	3	4
Balance	\$210	\$195	\$180	\$165	\$150

An examination of the values reveals that the account loses \$15 every month (the month increases by one and the balance decreases by 15). This information can be used to predict future values. To determine what the value will be in month 6, the pattern can be continued, and it can be concluded that the balance will be \$120. To determine which month the balance will be \$0, \$210 is divided by \$15 (since the balance decreases \$15 every month), resulting in month 14.

Similar to a table, a graph can display corresponding values of a linear relationship.

$y = 2x$
$40 = 2(20)$

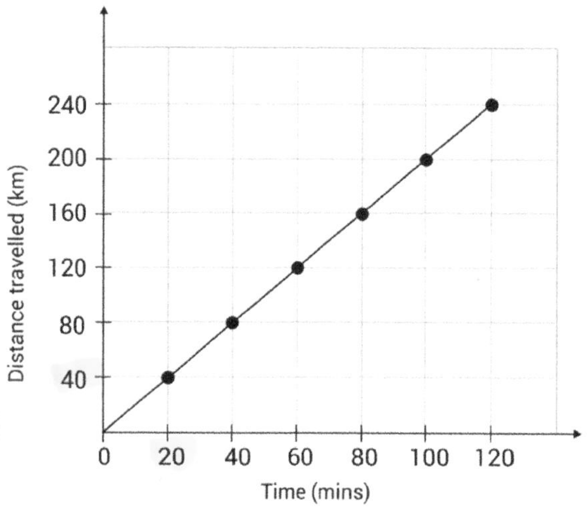

The graph above represents the relationship between distance traveled and time. To find the distance traveled in 80 minutes, the mark for 80 minutes is located at the bottom of the graph. By following this mark directly up on the graph, the corresponding point for 80 minutes is directly across from the 160 kilometer mark. This information indicates that the distance travelled in 80 minutes is 160 kilometers. To predict information not displayed on the graph, the way in which the variables change with respect to one another is determined. In this case, distance increases by 40 kilometers as time increases by 20 minutes. This information can be used to continue the data in the graph or convert the values to a table.

<u>Representing Proportional Relationships by Equations</u>

Linear relationships describe the way two quantities change with respect to each other. The relationship is defined as linear because a line is produced if all the sets of corresponding values are graphed on a coordinate grid. Linear relationships are proportional. A **linear equation** of the form $y = mx + b$ has two important quantities: m and b. The variable m represents the slope of the line, and the b indicates the y-intercept of the line. Slope is equivalent to the rate of change or the amount that y changes for a change in x. A positive value of m indicates a positive rate of change, while a negative value of m indicates a negative rate of change. The y-intercept, b, indicates the y-value, or the quantity of y, when x is equal to 0.

The two quantities, *m* and *b*, are especially meaningful when the linear equation models (or represents) an actual real-life situation. In many of these real-world applications, the *y*-intercept represents the initial amount of the quantity *y* because the *x*-variable would never be a negative quantity. For example, *x* often indicates time, and in many of these scenarios, the *x*-variable is never negative because time would not be negative.

Consider the following example. These two equations represent the cost, *C*, of gigabytes of data on a phone plan, *g*, at two different cellular phone companies:

$$C = 9g$$

$$C = 6g + 18$$

The first equation represents a scenario that shows the cost per gigabyte of data is $9. In this equation, *g* varies directly with *y*. There is no *y*-intercept, which means that there is no initial cost for a data plan with that company. The rate of change is 9, which is price per gigabyte of data. The second equation represents a scenario that has both an initial cost and a cost per gigabyte of data. The slope, which is 6, means that each gigabyte is $6. The *y*-intercept of 18 shows that there is an initial cost of a data plan with that company. Therefore, it makes sense to use the first company at $9 a gigabyte when only purchasing a small number of gigabytes of data. However, any large data needs would be cheaper by going with the second company because eventually that initial cost will be negligible. The exact tipping point can be determined by setting the equations equal to one another and solving for *g:*

$$9g = 6g + 18$$

$$3g = 18$$

$$g = 6$$

Therefore, both companies charge the same when 6 gigabytes of data are purchased. When fewer than 6 gigabytes are purchased, the first company is a better value because it doesn't charge the initial cost, and when more than 6 gigabytes of data are purchased, the second company is a better value because the rate per gigabyte is less.

Consider another example. Suppose a teacher is grading a test containing 20 questions with 5 points given for each correct answer, adding a curve of 10 points to each test. This linear relationship can be expressed as the equation $y = 5x + 10$ where *x* represents the number of correct answers and *y* represents the test score. To determine the score of a test with a given number of correct answers, the number of correct answers is substituted into the equation for *x* and evaluated. For example, for 10 correct answers, 10 is substituted for *x*:

$$y = 5(10) + 10 \rightarrow y = 60$$

Therefore, 10 correct answers will result in a score of 60. The number of correct answers needed to obtain a certain score can also be determined. To determine the number of correct answers needed to score a 90, 90 is substituted for *y* in the equation (*y* represents the test score) and solved:

$$90 = 5x + 10 \rightarrow 80 = 5x \rightarrow 16 = x$$

Therefore, 16 correct answers are needed to score a 90.

Explaining What a Point (x, y) on the Graph of a Proportional Relationship Means

A linear equation with two variables can be written given a point (ordered pair) and the slope or given two points on a line. An ordered pair gives a set of corresponding values for the two variables (*x* and *y*). As an example, for a scenario involving distance and time, it is given that the person traveled 112.5 miles in 2 ½ hours. Knowing that *x* represents time and *y* represents distance, this information can be written as the ordered pair (2.5, 112.5).

As mentioned, linear relationships describe the way two quantities change with respect to each other. The relationship is defined as linear because a line is produced if all the sets of corresponding values are graphed on a coordinate grid. When expressing the linear relationship as an equation, the equation is often written in the form $y = mx + b$ (**slope-intercept form**) where *m* and *b* are numerical values and *x* and *y* are variables (for example, $y = 5x + 10$). The slope is the coefficient of x, and the y-intercept is the constant value. The slope of the line containing the same two points is $m = \frac{y_2 - y_1}{x_2 - x_1}$ and is also equal to rise/run. Given a linear equation and the value of either variable (*x* or *y*), the value of the other variable can be determined.

The first step to writing a linear model is to identify what the variables represent. A **variable** represents an unknown quantity, and in the case of a linear equation, a specific relationship exists between the two variables (usually *x* and *y*). Within a given scenario, the variables are the two quantities that are changing. The variable *x* is considered the **independent variable** and represents the inputs of a function. The variable *y* is considered the **dependent variable** and represents the outputs of a function. For example, if a scenario describes distance traveled and time traveled, distance would be represented by *y* and time represented by *x*. The distance traveled depends on the time spent traveling (time is independent). If a scenario describes the cost of a cab ride and the distance traveled, the cost would be represented by *y* and the distance represented by *x*. The cost of a cab ride depends on the distance traveled.

As mentioned, the slope represents the rate of change. In linear models, because the slope is constant (a straight line), the rate of change is also constant. In the case of the cab ride, as the distance of the ride increases, so too does the cost in a directly proportional rate.

The slope of the graph of a line represents the rate of change between the variables of an equation. In the context of a real-world scenario, the slope will tell the way in which the unknown quantities (variables) change with respect to each other. A scenario involving distance and time might state that someone is traveling at a rate of 45 miles per hour. The slope of the linear model would be 45. A scenario involving the cost of a cab ride and distance traveled might state that the person is charged $3 for each mile. The slope of the linear model would be 3.

The *y*-intercept of a linear function is the value of *y* when $x = 0$ (the point where the line intercepts the *y*-axis on the graph of the equation). It is sometimes helpful to think of this as a "starting point" for a linear function. Suppose for the scenario about the cab ride that the person is told that the cab company charges a flat fee of $5 plus $3 for each mile. Before traveling any distance ($x = 0$), the cost is $5. The *y*-intercept for the linear model would be 5.

Using Proportional Relationships to Solve Multistep Ratio and Percent Problems

Questions dealing with percentages can be difficult when they are phrased as word problems. These word problems almost always come in three varieties. The first type will ask to find what percentage of some number will equal another number. The second asks to determine what number is some percentage of another given number. The third will ask what number another number is a given percentage of.

One of the most important parts of correctly answering percentage word problems is to identify the numerator and the denominator. This fraction can then be converted into a percentage, as described above.

The following word problem shows how to make this conversion:

A department store carries several different types of footwear. The store is currently selling 8 athletic shoes, 7 dress shoes, and 5 sandals. What percentage of the store's footwear are sandals?

First, calculate what serves as the 'whole', as this will be the denominator. How many total pieces of footwear does the store sell? The store sells 20 different types (8 athletic + 7 dress + 5 sandals).

Second, what footwear type is the question specifically asking about? Sandals. Thus, 5 is the numerator.

Third, the resultant fraction must be expressed as a percentage. The first two steps indicate that $\frac{5}{20}$ of the footwear pieces are sandals. This fraction must now be converted into a percentage:

$$\frac{5}{20} \times \frac{5}{5} = \frac{25}{100} = 25\%$$

The Number System

Adding, Subtracting, Multiplying, and Dividing Rational Numbers

Adding and Subtracting Rational Numbers and Representing Addition and Subtraction on a Number Line

Situations in Which Opposite Quantities Combine to Make 0

Aside from 0, numbers can be either positive or negative. The sign for a positive number is the plus sign or the + symbol, while the sign for a negative number is minus sign or the – symbol. If a number has no designation, then it's assumed to be positive. If a positive and negative number with the same absolute value are added, their sum is zero; in other words, they cancel each other out. Both 4 and -4 are four spaces from 0 on a number line because they both have an absolute value of 4. Adding 4 + -4 results in zero.

There are real-life situations in which these sorts of opposite quantities combine to make zero. Imagine in a football game your favorite team gains ten yards during a play. That can be represented by +10. On the next play, unfortunately, there's a tackle and they lose ten yards, which can be represented by -10. After the two plays, the net movement down the field is 0 yards.

Understanding Addition and Subtraction as Inverse Operations and Interpreting Sums of Rational Numbers by Describing Real-World Contexts

Addition and subtraction are **inverse operations**. Adding a number and then subtracting the same number will cancel each other out, resulting in the original number, and vice versa. For example, $8 + 7 - 7 = 8$ and $137 - 100 + 100 = 137$. Similarly, multiplication and division are inverse operations. Therefore, multiplying by a number and then dividing by the same number results in the original number, and vice versa. For example, $8 \times 2 \div 2 = 8$ and $12 \div 4 \times 4 = 12$. Inverse operations are used to work backwards to solve problems. In the case that 7 and a number add to 18, the inverse operation of subtraction is used to find the unknown value $(18 - 7 = 11)$. If a school's entire 4th grade was divided evenly into 3 classes each with 22 students, the inverse operation of multiplication is used to determine the total students in the grade $(22 \times 3 = 66)$.

Additional scenarios involving inverse operations are included in the tables below.

There are a variety of real-world situations in which one or more of the operators is used to solve a problem. The tables below display the most common scenarios.

	Unknown Result	Unknown Change	Unknown Start
Adding to	5 students were in class. 4 more students arrived. How many students are in class? $5 + 4 = ?$	8 students were in class. More students arrived late. There are now 18 students in class. How many students arrived late? $8 + ? = 18$ Solved by inverse operations $18 - 8 = ?$	Some students were in class early. 11 more students arrived. There are now 17 students in class. How many students were in class early? $? + 11 = 17$ Solved by inverse operations $17 - 11 = ?$
Taking from	15 students were in class. 5 students left class. How many students are in class now? $15 - 5 = ?$	12 students were in class. Some students left class. There are now 8 students in class. How many students left class? $12 - ? = 8$ Solved by inverse operations $8 + ? = 12 \rightarrow 12 - 8 = ?$	Some students were in class. 3 students left class. Then there were 13 students in class. How many students were in class before? $? - 3 = 13$ Solved by inverse operations $13 + 3 = ?$

	Unknown Total	Unknown Addends (Both)	Unknown Addends (One)
Putting together/ taking apart	The homework assignment is 10 addition problems and 8 subtraction problems. How many problems are in the homework assignment? $10 + 8 = ?$	Bobby has $9. How much can Bobby spend on candy and how much can Bobby spend on toys? $9 = ? + ?$	Bobby has 12 pairs of pants. 5 pairs of pants are shorts, and the rest are long. How many pairs of long pants does he have? $12 = 5 + ?$ Solved by inverse operations $12 - 5 = ?$

	Unknown Difference	Unknown Larger Value	Unknown Smaller Value
Comparing	Bobby has 5 toys. Tommy has 8 toys. How many more toys does Tommy have than Bobby? $5 + ? = 8$ Solved by inverse operations $8 - 5 = ?$ Bobby has \$6. Tommy has \$10. How many fewer dollars does Bobby have than Tommy? $10 - 6 = ?$	Tommy has 2 more toys than Bobby. Bobby has 4 toys. How many toys does Tommy have? $2 + 4 = ?$ Bobby has 3 fewer dollars than Tommy. Bobby has \$8. How many dollars does Tommy have? $? - 3 = 8$ Solved by inverse operations $8 + 3 = ?$	Tommy has 6 more toys than Bobby. Tommy has 10 toys. How many toys does Bobby have? $? + 6 = 10$ Solved by inverse operations $10 - 6 = ?$ Bobby has \$5 less than Tommy. Tommy has \$9. How many dollars does Bobby have? $9 - 5 = ?$

Some problems require adding positive and negative numbers or subtracting positive and negative numbers. Adding a negative number to a positive one can be thought of a reducing or subtracting from the positive number, and the result should be less than the original positive number. For example, adding 8 and -3 is the same is subtracting 3 from 8; the result is 5. This can be visualized by imagining that the positive number (8) represents 8 apples that a student has in her basket. The negative number (-3) indicates the number of apples she is in debt or owes to her friend. In order to pay off her debt and "settle the score," she essentially is in possession of three fewer apples than in her basket (8 – 3 = 5), so she actually has five apples that are hers to keep. Should the negative addend be of higher magnitude than the positive addend (for example -9 + 3), the result will be negative, but "less negative" or closer to zero than the large negative number. This is because adding a positive value, even if relatively smaller, to a negative value, reduces the magnitude of the negative in the total. Considering the apple example again, if the girl owed 9 apples to her friend (-9) but she picked 3 (+3) off a tree and gave them to her friend, she now would only owe him six apples (-6), which reduced her debt burden (her negative number of apples) by three.

Subtracting positive and negative numbers works the same way with one key distinction: subtracting a negative number from a negative number yields a "less negative" or more positive result because again, this can be considered as removing or alleviating some debt. For example, if the student with the apples owed 5 apples to her friend, she essentially has -5 applies. If her mom gives that friend 10 apples on behalf of the girl, she now has removed the need to pay back the 5 apples and surpassed neutral (no net apples owed) and now her friend owes *her* five apples (+5). Stated mathematically -5 + -10 = +5.

When subtracting integers and negative rational numbers, one has to change the problem to adding the opposite and then apply the rules of addition.

Understanding Subtraction as Adding the Additive Inverse and Understanding Absolute Value

Rational numbers are all numbers that can be written as a fraction using integers. A **fraction** is written as $\frac{x}{y}$ and represents the quotient of *x* being divided by *y*. More practically, it means dividing the whole into *y* equal parts, then taking *x* of those parts.

Examples of rational numbers include $\frac{1}{2}$ and $\frac{5}{4}$. The number on the top is called the **numerator,** and the number on the bottom is called the **denominator**. Because every integer can be written as a fraction with a denominator of 1, (e.g. $\frac{3}{1} = 3$), every integer is also a rational number.

As mentioned, when adding integers and negative rational numbers, there are some basic rules to determine if the solution is negative or positive:

- Adding two positive numbers results in a positive value: $3.3 + 4.8 = 8.1$.
- Adding two negative numbers results in a negative number: $(-8) + (-6) = -14$.

Adding one positive and one negative number requires taking the absolute values and finding the difference between them. Then, the sign of the number that has the higher absolute value for the final solution is used.

For example, $(-9) + 11$, has a difference of absolute values of 2. The final solution is 2 because 11 has the higher absolute value. Another example is $9 + (-11)$, which has a difference of absolute values of 2. The final solution is -2 because 11 has the higher absolute value.

When subtracting integers and negative rational numbers, one has to change the problem to adding the opposite and then apply the rules of addition.

- Subtracting two positive numbers is the same as adding one positive and one negative number. For example, $4.9 - 7.1$ is the same as $4.9 + (-7.1)$. The solution is -2.2 since the absolute value of -7.1 is greater. Another example is $8.5 - 6.4$, which is the same as $8.5 + (-6.4)$. The solution is 2.1 since the absolute value of 8.5 is greater.

- Subtracting a positive number from a negative number results in negative value. For example, $(-12) - 7$ is the same as $(-12) + (-7)$ with a solution of -19.

- Subtracting a negative number from a positive number results in a positive value. For example, $12 - (-7)$ is the same as $12 + 7$ with a solution of 19.

Absolute Value

Both positive and negative numbers are valued according to their distance from 0. Both +3 and -3 can be considered using the following number line:

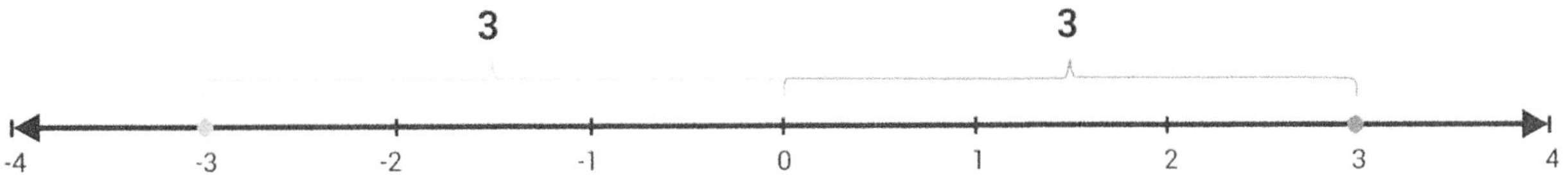

Both 3 and -3 are three spaces from 0. The distance from 0 is called its **absolute value**. Thus, both -3 and 3 have an absolute value of 3 since they're both three spaces away from 0.

An absolute number is written by placing | | around the number. So, $|3|$ and $|-3|$ both equal 3, as that's their common absolute value.

Applying Properties of Operations as Strategies to Add and Subtract Rational Numbers
Properties of operations exist that make calculations easier and solve problems for missing values. The following table summarizes commonly used properties of real numbers for addition.

Property	**Addition**
Commutative	$a + b = b + a$
Associative	$(a + b) + c = a + (b + c)$
Identity	$a + 0 = a;\ 0 + a = a$
Inverse	$a + (-a) = 0$

Addition holds the **commutative property**, which means that numbers in an addition equation can be switched without altering the result. The formula for the commutative property is $a + b = b + a$. Let's look at a few examples to see how the commutative property works:

$$3 + 4 = 4 + 3 = 7$$

$$12 + 8 = 8 + 12 = 20$$

Addition also holds the **associative property**, which means that the groups of numbers don't matter in an addition problem. The formula for the associative property is $(a + b) + c = a + (b + c)$. Here are some examples of the Associative Property at work:

$$(6 + 14) + 10 = 6 + (14 + 10) = 30$$

$$8 + (2 + 25) = (8 + 2) + 25 = 35$$

The commutative and associative properties are useful for performing calculations. For example, $(47 + 25) + 3$ is equivalent to $(47 + 3) + 25$, which is easier to calculate.

The **identity property of addition** states that adding zero to any number does not change its value. The **inverse property of addition** states that the sum of a number and its opposite equals zero. Opposites are numbers that are the same with different signs (ex. 5 and -5; $-\frac{1}{2}$ and $\frac{1}{2}$). Inverse properties are useful for canceling quantities to find missing values. For example, $a + 7 = 12$ is solved by adding the inverse of 7(-7) to both sides in order to isolate *a*.

Unlike addition, subtraction doesn't follow the commutative or associative properties. The order and grouping in subtraction impacts the result.

$$22 - 7 \neq 7 - 22$$

$$(10 - 5) - 2 \neq 10 - (5 - 2)$$

Operations with Fractions

Multiplying and Dividing Rational Numbers

Multiplying Rational Numbers and Interpreting Products by Describing Real-World Contexts
There are a variety of real-world situations in which multiplication and division are used to solve problems. The tables below display some of the most common scenarios.

	Unknown Product	**Unknown Group Size**	**Unknown Number of Groups**
Equal groups	There are 5 students, and each student has 4 pieces of candy. How many pieces of candy are there in all? $5 \times 4 = ?$	14 pieces of candy are shared equally by 7 students. How many pieces of candy does each student have? $7 \times ? = 14$ Solved by inverse operations $14 \div 7 = ?$	If 18 pieces of candy are to be given out 3 to each student, how many students will get candy? $? \times 3 = 18$ Solved by inverse operations $18 \div 3 = ?$

	Unknown Product	**Unknown Factor**	**Unknown Factor**
Arrays	There are 5 rows of students with 3 students in each row. How many students are there? $5 \times 3 = ?$	If 16 students are arranged into 4 equal rows, how many students will be in each row? $4 \times ? = 16$ Solved by inverse operations $16 \div 4 = ?$	If 24 students are arranged into an array with 6 columns, how many rows are there? $? \times 6 = 24$ Solved by inverse operations $24 \div 6 = ?$

	Larger Unknown	**Smaller Unknown**	**Multiplier Unknown**
Comparing	A small popcorn costs \$1.50. A large popcorn costs 3 times as much as a small popcorn. How much does a large popcorn cost? $1.50 \times 3 = ?$	A large soda costs \$6 and that is 2 times as much as a small soda costs. How much does a small soda cost? $2 \times ? = 6$ Solved by inverse operations $6 \div 2 = ?$	A large pretzel costs \$3 and a small pretzel costs \$2. How many times as much does the large pretzel cost as the small pretzel? $? \times 2 = 3$ Solved by inverse operations $3 \div 2 = ?$

Dividing Integers and Interpreting Quotients of Rational Numbers in Real-World Contexts
Rational numbers include all numbers that can be expressed as a fraction; in other words, rational numbers encompass all integers and all numbers with terminating or repeating decimals. That is, any rational number either will have a countable number of nonzero digits or will end with an ellipses or a bar (3.6666... or $3.\overline{6}$) to depict repeating decimal digits.

Some examples of rational numbers include 12, -3.54, $110.\overline{256}$, $\frac{-35}{10}$, and $4.\overline{7}$.

Suppose Maggie is reading a book with 149 pages. Maggie has read 10 pages per day and has 9 pages left to read. How many days has Maggie been reading?

The solution can be demonstrated using an equation or a strip diagram to set up the problem.

Equation method:

Let *d* be the number of days Maggie has been reading. Set up the basic multiplication of 10 pages times *d* days, along with the additional 9 pages left to read. This should be set equal to the total number of pages, 149:

$$10d + 9 = 149$$ Solve for *d*:

$$\begin{aligned} 10d + 9 &= 149 \\ \underline{-9} &= \underline{-9} \\ 10d &= 140 \\ 10d/10 &= 140/10 \end{aligned}$$

$$d = 14\ days$$

A number strip would display similar information visually:

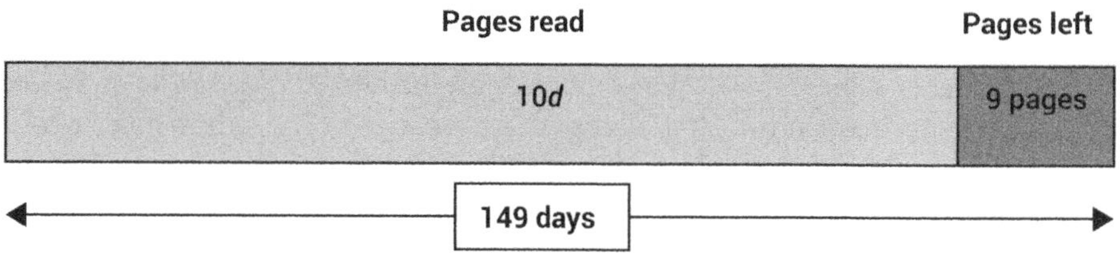

Both methods assist in the organization of the information given in the initial problem and both help to determine that there will be more than one step involved in solving the problem. In most cases, it will not matter in which order the steps are executed; the setup is the important factor.

Applying Properties of Operations as Strategies to Multiply and Divide Rational Numbers

As mentioned, properties of operations exist that make calculations easier and solve problems for missing values. The following table summarizes commonly used properties of real numbers for multiplication.

Property	Multiplication
Commutative	$a \times b = b \times a$
Associative	$(a \times b) \times c = a \times (bc)$
Identity	$a \times 1 = a;\ 1 \times a = a$
Inverse	$a \times \frac{1}{a} = 1;\ a \neq 0$
Distributive	$a(b + c) = ab + ac$

Like addition, multiplication holds the commutative and associative properties. The **commutative property of multiplication** states that the order in which numbers are multiplied does not change the product.

$$2 \times 4 = 4 \times 2$$

$$2 \times 5 = 5 \times 20$$

The **associative property** of multiplication state that the grouping of numbers being multiplied does not change the product.

$$(2 \times 4) \times 10 = 2 \times (4 \times 10)$$

$$3 \times (7 \times 4) = (3 \times 7) \times 4$$

The **identity property of multiplication** states that multiplying a number by 1 does not change its value. The **inverse property of multiplication** states that the product of a number (other than 0) and its reciprocal equals 1. **Reciprocal numbers** have numerators and denominators that are inverted (ex. $\frac{2}{5}$ and $\frac{5}{2}$).

Multiplication also follows the **distributive property**, which allows the multiplication to be distributed through parentheses. The distributive property states that multiplying a sum (or difference) by a number produces the same result as multiplying each value in the sum (or difference) by the number and adding (or subtracting) the products. The formula for distribution is $a \times (b + c) = ab + ac$. This will become clearer after some examples:

$$5 \times (3 + 6) = (5 \times 3) + (5 \times 6) = 45$$

$$4 \times (10 - 5) = (4 \times 10) - (4 \times 5) = 20$$

Consider the following scenario: You are buying three tickets for a baseball game. Each ticket costs $18. You are also charged a fee of $2 per ticket for purchasing the tickets online. The cost is calculated: $3 \times 18 + 3 \times 2$. Using the distributive property, the cost can also be calculated $3(18 + 2)$.

Division doesn't follow any of the commutative, associative, or distributive properties.

$$15 \div 3 \neq 3 \div 15$$

$$(30 \div 3) \div 5 \neq 30 \div (3 \div 5)$$

$$100 \div (20 + 10) \neq (100 \div 20) + (100 \div 10)$$

For multiplication and division of integers and rational numbers, if both numbers are positive or both numbers are negative, the result is a positive value. For example, $(-1.7)(-4)$ has a solution of 6.8 since both numbers are negative values.

If one number is positive and another number is negative, the result is a negative value. For example, $(-15) \div 5$ has a solution of -3 since there is one negative number.

<u>Converting a Rational Number to a Decimal Using Long Division</u>

The **decimal system** is a way of writing out numbers that uses ten different numerals: 0, 1, 2, 3, 4, 5, 6, 7, 8, and 9. This is also called a "base ten" or "base 10" system. Other bases are also used. For example, computers work with a base of 2. This means they only use the numerals 0 and 1.

The **decimal place** denotes how far to the right of the decimal point a numeral is. The first digit to the right of the decimal point is in the *tenths* place. The next is the **hundredths**. The third is the **thousandths**.

So, 3.142 has a 1 in the tenths place, a 4 in the hundredths place, and a 2 in the thousandths place.

The **decimal point** is a period used to separate the **ones** place from the **tenths** place when writing out a number as a decimal.

A **decimal number** is a number written out with a decimal point instead of as a fraction, for example, 1.25 instead of $\frac{5}{4}$. Thus, a rational number can be converted to a decimal using long division. For example, the rational number $\frac{5}{4}$ can be converted to a decimal by dividing the numerator by the denominator. Depending on the situation, it can sometimes be easier to work with fractions and sometimes easier to work with decimal numbers.

A decimal number is **terminating** if it stops at some point. It is called **repeating** if it never stops, but repeats a pattern over and over. It is important to note that every rational number can be written as a terminating decimal or as a repeating decimal.

Solving Real-World and Mathematical Problems Involving the Four Operations with Rational Numbers

Again, a **rational number** is any number that can be written in the form of a ratio or fraction. Integers can be written as fractions with a denominator of 1 ($5 = \frac{5}{1}$; $-342 = \frac{-342}{1}$; etc.). Decimals that terminate and/or repeat can also be written as fractions ($47 = \frac{47}{100}$; $.\overline{33} = \frac{1}{3}$).

When adding or subtracting fractions, the numbers must have the same denominators. In these cases, numerators are added or subtracted, and denominators are kept the same. For example, $\frac{2}{7} + \frac{3}{7} = \frac{5}{7}$ and $\frac{4}{5} - \frac{3}{5} = \frac{1}{5}$. If the fractions to be added or subtracted do not have the same denominator, a common denominator must be found. This is accomplished by changing one or both fractions to a different but equivalent fraction. Consider the example $\frac{1}{6} + \frac{4}{9}$. First, a common denominator must be found. One method is to find the least common multiple (LCM) of the denominators 6 and 9. This is the lowest number that both 6 and 9 will divide into evenly. In this case the LCM is 18. Both fractions should be changed to equivalent fractions with a denominator of 18. To obtain the numerator of the new fraction, the old numerator is multiplied by the same number by which the old denominator is multiplied. For the fraction $\frac{1}{6}$, 6 multiplied by 3 will produce a denominator of 18. Therefore, the numerator is multiplied by 3 to produce the new numerator $\left(\frac{1\times3}{6\times3} = \frac{3}{18}\right)$. For the fraction $\frac{4}{9}$, multiplying both the numerator and denominator by 2 produces $\frac{8}{18}$. Since the two new fractions have common denominators, they can be added $\left(\frac{3}{18} + \frac{8}{18} = \frac{11}{18}\right)$.

When multiplying or dividing rational numbers, these numbers may be converted to fractions and multiplied or divided accordingly. When multiplying fractions, all numerators are multiplied by each other and all denominators are multiplied by each other. For example, $\frac{1}{3} \times \frac{6}{5} = \frac{1\times6}{3\times5} = \frac{6}{15}$ and $\frac{-1}{2} \times \frac{3}{1} \times \frac{11}{100} = \frac{-1\times3\times11}{2\times1\times100} = \frac{-33}{200}$. When dividing fractions, the problem is converted by multiplying by the reciprocal of the divisor. This is done by changing division to multiplication and "flipping" the second fraction, or divisor. For example, $\frac{1}{2} \div \frac{3}{5} \rightarrow \frac{1}{2} \times \frac{5}{3}$ and $\frac{5}{1} \div \frac{1}{3} \rightarrow \frac{5}{1} \times \frac{3}{1}$. To complete the problem, the rules for multiplying fractions should be followed.

Note that when adding, subtracting, multiplying, and dividing mixed numbers (ex. $4\frac{1}{2}$), it is easiest to convert these to improper fractions (larger numerator than denominator). To do so, the denominator is kept the same. To obtain the numerator, the whole number is multiplied by the denominator and added to the numerator. For example, $4\frac{1}{2} = \frac{9}{2}$ and $7\frac{2}{3} = \frac{23}{3}$. Also, note that answers involving fractions should be converted to the simplest form.

Word and story problems should be presented in many different ways. Word problems present the opportunity to related mathematical concepts from the classroom into real-world situations. These types of problems are situations in which some parts of the problem are known and at least one part is unknown.

There are three types of instances in which something can be unknown: the starting point, the modification, or the final result can all be missing from the provided information.

- For an addition problem, the **modification** is the quantity of the new amount added to the starting point.

- For a subtraction problem, the **modification** is the quantity taken away from the starting point.

Keywords in the word problems can signal what type of operation needs to be used to solve the problem. Words such as *total, increased, combined*, and *more* indicate that addition is needed. Words such as *difference, decreased,* and *minus* indicate that subtraction is needed.

Regarding addition, the given equation is $3 + 7 = 10$.

The number 3 is the starting point. 7 is the modification, and 10 is the result from adding a new amount to the starting point. Different word problems can arise from this same equation, depending on which value is the unknown. For example, here are three problems:

- *If a student had three pencils and was given seven more, how many would he have in total?*

- *If a student had three pencils and a student gave him more so that he had ten in total, how many were given to him?*

- *A student was given seven pencils so that he had ten in total. How many did he start with?*

All three problems involve the same equation, and determining which part of the equation is missing is the key to solving each word problem. The missing answers would be 10, 7, and 3, respectively.

In terms of subtraction, the same three scenarios can occur. The given equation is $6 - 4 = 2$.

The number 6 is the starting point. 4 is the modification, and 2 is the new amount that is the result from taking away an amount from the starting point. Again, different types of word problems can arise from this equation. For example, here are three possible problems:

- *If a student had six quarters and four were taken away, how many would be left over?*

- *If a student had six quarters, purchased a pencil, and had two quarters left over, how many quarters did she pay with?*

- *If a student paid for a pencil with four quarters and had two quarters left over, how many did she have to start with?*

The three question types follow the structure of the addition word problems, and determining whether the starting point, the modification, or the final result is missing is the goal in solving the problem. The missing answers would be 2, 4, and 6, respectively.

The three addition problems and the three subtraction word problems can be solved by using a picture, a number line, or an algebraic equation. If an equation is used, a question mark can be utilized to represent the unknown quantity. For example, $6 - 4 =?$ can be written to show that the missing value is the result. Using equation form visually indicates what portion of the addition or subtraction problem is the missing value.

Similar instances can be seen in word problems involving multiplication and division. Key words within a multiplication problem involve *times, product, doubled*, and *tripled.* Key words within a division problem involve *split, quotient, divided, shared, groups*, and *half*. Like addition and subtraction, multiplication and division problems also have three different types of missing values.

Multiplication consists of a specific number of groups having the same size, the quantity of items within each group, and the total quantity within all groups. Therefore, each one of these amounts can be the missing value.

For example, the given equation is $5 \times 3 = 15$.

5 and 3 are interchangeable, so either amount can be the number of groups or the quantity of items within each group. 15 is the total number of items. Again, different types of word problems can arise from this equation. For example, here are three problems:

- If a classroom is serving 5 different types of apples for lunch and has three apples of each type, how many total apples are there to give to the students?
- If a classroom has 15 apples of 5 different types, how many of each type are there?
- If a classroom has 15 apples with 3 of each type, how many types are there to choose from?

Each question involves using the same equation to solve, and it is imperative to decide which part of the equation is the missing value. The answers to the problems are 15, 3, and 5, respectively.

Similar to multiplication, division problems involve a total amount, a number of groups having the same size, and a number of items within each group. The difference between multiplication and division is that the starting point is the total amount, which then gets divided into equal quantities.

For example, the equation is $15 \div 5 = 3$.

15 is the total number of items, which is being divided into 5 different groups. In order to do so, 3 items go into each group. Also, 5 and 3 are interchangeable, so the 15 items could be divided into 3 groups of 5 items each. Therefore, different types of word problems can arise from this equation depending on which value is unknown. For example, here are three types of problems:

- *A student needs 48 pieces of chalk. If there are 8 pieces in each box, how many boxes should he buy?*

- *A student has 48 pieces of chalk. If each box has 6 pieces in it, how many boxes did he buy?*

- *A student has partitioned all of his chalk into 8 piles, with 6 pieces in each pile. How many pieces does he have in total?*

Each one of these questions involves the same equation, and the third question can easily utilize the multiplication equation $8 \times 6 = ?$ instead of division. The answers are 6, 8, and 48, respectively.

Calculations relating to real-world expenses rarely have whole number solutions. It is good practice to be able to make one-step calculations that model real-world situations. For example, Amber spends $54 on pet food in a month. If there are no increases in price for the next 11 months, how much will Amber spend on pet food during those 11 months?

Set up the multiplication problem to see how much $54 of food times 11 is:

$$\begin{array}{r} 54 \\ \times\ \underline{11} \\ 54 \\ +\underline{540} \\ 594 \end{array}$$

Amber would spend $594 on pet food over the next 11 months.

As another example, what if Amber had $650; how many months would this last her for pet food expenses if they were $54 per month?

Set the problem up as a division problem to see how many times 54 could divide into 650. The answer will give the number of complete months the expenses could be covered.

$$\begin{array}{r} 12 \\ 54\overline{)650} \\ -\ 54 \\ \hline 110 \\ -\ 108 \\ \hline 2 \end{array}$$

The remainder does not represent a full month of expenses. Therefore, Amber's $650 would last for a full 12 months of pet food expenses.

Typical representations of real-world situations involve organizing data and systematically applying a calculation to a set of numbers, so as to calculate a systematic output.

For example, a bicycle store is taking inventory of the number of tires on their bikes. Create a table in which to display the possible outputs there could be for two-wheeled bikes and a column to display how these outputs are calculated. How was this table was designed, and how were the calculations that represent the data determined?

Create a table with three categories for the input, the calculation, and the output. The following represent possible products:

Number of Bikes (input)	Calculation of Wheels	Total Number of Wheels (output)
1	1×2	2
2	2×2	4
3	3×2	6
5	5×2	10
10	10×2	20

Or the table could be organized as follows:

Number of Bikes (input)	1	2	3	5	10
Calculation of Wheels	1×2	2×2	3×2	5×2	10×2
Total Number of Wheels (output)	2	4	6	10	20

Be able to describe the process used to design the table: the categories for input, calculation, and output. Note that the calculations encompass the idea that for every bike there are two wheels. Thus, the multiplication of the number of bikes by the number of wheels (a constant of 2) results in the output for each permutation.

Work word problems are examples of people working together in a situation that uses fractions. One painter can paint a designated room in 6 hours, and a second painter can paint the same room in 5 hours. How long will it take them to paint the room if they work together?

The first painter paints $\frac{1}{6}$ of the room in an hour, and the second painter paints $\frac{1}{5}$ of the room in an hour.

Together, they can paint $\frac{1}{x}$ of the room in an hour. The equation is the sum of the painter's rate equal to the total job or $\frac{1}{6} + \frac{1}{5} = \frac{1}{x}$.

The equation can be solved by multiplying all terms by a common denominator of 30*x* with a result of $5x + 6x = 30$.

The left side can be added together to get 11*x*, and then divide by 11 for a solution of $\frac{30}{11}$ or about 2.73 hours.

Expressions & Equations

Using Properties of Operations to Generate Equivalent Expressions

Applying Properties of Operations as Strategies to Add, Subtract, Factor, and Expand Linear Expressions

A fraction, or ratio, wherein each part is a polynomial, defines **rational expressions**. Some examples include $\frac{2x+6}{x}$, $\frac{1}{x^2-4x+8}$, and $\frac{z^2}{x+5}$. Exponents on the variables are restricted to whole numbers, which means roots and negative exponents are not included in rational expressions.

Rational expressions can be transformed by factoring. For example, the expression $\frac{x^2-5x+6}{(x-3)}$ can be rewritten by factoring the numerator to obtain:

$$\frac{(x-3)(x-2)}{(x-3)}$$

Therefore, the common binomial $(x-3)$ can cancel so that the simplified expression is:

$$\frac{(x-2)}{1} = (x-2)$$

Additionally, other rational expressions can be rewritten to take on different forms. Some may be factorable in themselves, while others can be transformed through arithmetic operations. Rational expressions are closed under addition, subtraction, multiplication, and division by a nonzero expression. **Closed** means that if any one of these operations is performed on a rational expression, the result will still be a rational expression. The set of all real numbers is another example of a set closed under all four operations.

Adding and subtracting rational expressions is based on the same concepts as adding and subtracting simple fractions. For both concepts, the denominators must be the same for the operation to take place. For example, here are two rational expressions:

$$\frac{x^3-4}{(x-3)} + \frac{x+8}{(x-3)}$$

Since the denominators are both $(x-3)$, the numerators can be combined by collecting like terms to form:

$$\frac{x^3+x+4}{(x-3)}$$

If the denominators are different, they need to be made common (the same) by using the Least Common Denominator (LCD). Each denominator needs to be factored, and the LCD contains each factor that appears in any one denominator the greatest number of times it appears in any denominator. The original expressions need to be multiplied times a form of 1, which will turn each denominator into the

LCD. This process is like adding fractions with unlike denominators. It is also important when working with rational expressions to define what value of the variable makes the denominator zero. For this particular value, the expression is undefined.

Multiplication of rational expressions is performed like multiplication of fractions. The numerators are multiplied; then, the denominators are multiplied. The final fraction is then simplified. The expressions are simplified by factoring and cancelling out common terms. In the following example, the numerator of the second expression can be factored first to simplify the expression before multiplying:

$$\frac{x^2}{(x-4)} \times \frac{x^2 - x - 12}{2}$$

$$\frac{x^2}{(x-4)} \times \frac{(x-4)(x+3)}{2}$$

The $(x - 4)$ on the top and bottom cancel out:

$$\frac{x^2}{1} \times \frac{(x+3)}{2}$$

Then multiplication is performed, resulting in:

$$\frac{x^3 + 3x^2}{2}$$

Dividing rational expressions is similar to the division of fractions, where division turns into multiplying by a reciprocal. So the following expression can be rewritten as a multiplication problem:

$$\frac{x^2 - 3x + 7}{x - 4} \div \frac{x^2 - 5x + 3}{x - 4}$$

$$\frac{x^2 - 3x + 7}{x - 4} \times \frac{x - 4}{x^2 - 5x + 3}$$

The $x - 4$ cancels out, leaving:

$$\frac{x^2 - 3x + 7}{x^2 - 5x + 3}$$

The final answers should always be completely simplified. If a function is composed of a rational expression, the zeros of the graph can be found from setting the polynomial in the numerator as equal to zero and solving. The values that make the denominator equal to zero will either exist on the graph as a hole or a vertical asymptote.

FOIL Method

FOIL is a technique for generating polynomials through the multiplication of binomials. A **polynomial** is an expression of multiple variables (for example, x, y, z) in at least three terms involving only the four basic operations and exponents. FOIL is an acronym for First, Outer, Inner, and Last. "First" represents the multiplication of the terms appearing first in the binomials. "Outer" means multiplying the outermost terms. "Inner" means multiplying the terms inside. "Last" means multiplying the last terms of each binomial.

After completing FOIL and solving the operations, **like terms** are combined. To identify like terms, test takers should look for terms with the same variable and the same exponent.

For example, in $4x^2 - x^2 + 15x + 2x^2 - 8$, the $4x^2, -x^2$, and $2x^2$ are all like terms because they have the variable (x) and exponent (2). Thus, after combining the like terms, the polynomial has been simplified to:

$$5x^2 + 15x - 8$$

The purpose of FOIL is to simplify an equation involving multiple variables and operations. Although it sounds complicated, working through some examples will provide some clarity:

1. Simplify $(x + 10)(x + 4) =$

$(x \times x)$	+	$(x \times 4)$	+	$(10 \times x)$	+	(10×4)
First		Outer		Inner		Last

After multiplying these binomials, it's time to solve the operations and combine like terms. Thus, the expression becomes: $x^2 + 4x + 10x + 40 = 2x^2 + 14x + 40$.

2. Simplify $2x\ (4x^3 - 7y^2 + 3x^2 + 4)$

Here, a monomial ($2x$) is multiplied into a polynomial $(4x^3 - 7y^2 + 3x^2 + 4)$. Using the distributive property, the monomial gets multiplied by each term in the polynomial. This becomes:

$$2x(4x^3) - 2x(7y^2) + 2x(3x^2) + 2x(4)$$

Now, each monomial is simplified, starting with the coefficients:

$$(2 \times 4)(x \times x^3) - (2 \times 7)(x \times y^2) + (2 \times 3)(x \times x^2) + (2 \times 4)(x)$$

When multiplying powers with the same base, their exponents are added. Remember, a variable with no listed exponent has an exponent of 1, and exponents of distinct variables cannot be combined. This produces the answer:

$$8x^{1+3} - 14xy^2 + 6x^{1+2} + 8x = 8x^4 - 14xy^2 + 6x^3 + 8x$$

3. Simplify $(8x^{10}y^2z^4) \div (4x^2y^4z^7)$

The first step is to divide the coefficients of the first two polynomials: $8 \div 4 = 2$. The second step is to divide exponents with the same variable, which requires subtracting the exponents. This results in:

$$2(x^{10-2}y^{2-4}z^{4-7}) = 2x^8y^{-2}z^{-3}$$

However, the most simplified answer should include only positive exponents. Thus, $y^{-2}z^{-3}$ needs to be converted into fractions, respectively $\frac{1}{y^2}$ and $\frac{1}{z^3}$. Since the $2x^8$ has a positive exponent, it is placed in the numerator, and $\frac{1}{y^2}$ and $\frac{1}{z^3}$ are combined into the denominator, leaving $\frac{2x^8}{y^2z^3}$ as the final answer.

Rewriting Expressions to Aid the Problem-Solving Process

Algebraic expressions are made up of numbers, variables, and combinations of the two, using mathematical operations. Expressions can be rewritten based on their factors. For example, the expression $6x + 4$ can be rewritten as $2(3x + 2)$ because 2 is a factor of both $6x$ and 4. More complex expressions can also be rewritten based on their factors. The expression $x^4 - 16$ can be rewritten as $(x^2 - 4)(x^2 + 4)$. This is a different type of factoring, where a difference of squares is factored into a sum and difference of the same two terms. With some expressions, the factoring process is simple and only leads to a different way to represent the expression. With others, factoring and rewriting the expression leads to more information about the given problem.

In the following quadratic equation, factoring the binomial leads to finding the zeros of the function:

$$x^2 - 5x + 6 = y$$

This equations factors into $(x - 3)(x - 2) = y$, where 2 and 3 are found to be the zeros of the function when y is set equal to zero. The zeros of any function are the x-values where the graph of the function on the coordinate plane crosses the x-axis.

Factoring an equation is a simple way to rewrite the equation and find the zeros, but factoring is not possible for every quadratic. Completing the square is one way to find zeros when factoring is not an option. The following equation cannot be factored: $x^2 + 10x - 9 = 0$. The first step in this method is to move the constant to the right side of the equation, making it $x^2 + 10x = 9$. Then, the coefficient of x is divided by 2 and squared. This number is then added to both sides of the equation, to make the equation still true. For this example, $\left(\frac{10}{2}\right)^2 = 25$ is added to both sides of the equation to obtain:

$$x^2 + 10x + 25 = 9 + 25$$

This expression simplifies to $x^2 + 10x + 25 = 34$, which can then be factored into $(x + 5)^2 = 34$. Solving for *x* then involves taking the square root of both sides and subtracting 5. This leads to two zeros of the function:

$$x = \pm\sqrt{34} - 5$$

Depending on the type of answer the question seeks, a calculator may be used to find exact numbers.

Given a quadratic equation in standard form— $ax^2 + bx + c = 0$—the sign of a tells whether the function has a minimum value or a maximum value. If $a > 0$, the graph opens up and has a minimum value. If $a < 0$, the graph opens down and has a maximum value. Depending on the way the quadratic equation is written, multiplication may need to occur before a max/min value is determined.

Exponential expressions can also be rewritten, just as quadratic equations. Properties of exponents must be understood. Multiplying two exponential expressions with the same base involves adding the exponents:

$$a^m a^n = a^{m+n}$$

Dividing two exponential expressions with the same base involves subtracting the exponents:

$$\frac{a^m}{a^n} = a^{m-n}$$

Raising an exponential expression to another exponent includes multiplying the exponents:

$$(a^m)^n = a^{mn}$$

The zero power always gives a value of 1: $a^0 = 1$. Raising either a product or a fraction to a power involves distributing that power:

$$(ab)^m = a^m b^m \text{ and } \left(\frac{a}{b}\right)^m = \frac{a^m}{b^m}$$

Finally, raising a number to a negative exponent is equivalent to the reciprocal including the positive exponent:

$$a^{-m} = \frac{1}{a^m}$$

Solving Real-Life and Mathematical Problems Using Numerical and Algebraic Expressions and Equations

Solving Multi-Step Problems with Rational Numbers and Assessing the Reasonableness of Answers

Exponents are shorthand for longer multiplications or divisions. The exponent is written to the upper right of a number. In the expression 2^3, the **exponent** is 3. The number with the exponent is called the **base.**

When the exponent is a whole number, it means to multiply the base by itself as many times as the number in the exponent. So, $2^3 = 2 \times 2 \times 2 = 8$.

If the exponent is a negative number, it means to take the reciprocal of the positive exponent:

$$2^{-3} = \frac{1}{2^3} = \frac{1}{8}$$

When the exponent is 0, the result is always 1: $2^0 = 1, 5^0 = 1$, and so on.

When the exponent is 2, the number is **squared,** and when the exponent is 3, it is **cubed.**

When working with longer expressions, parentheses are used to show the order in which the operations should be performed. Operations inside the parentheses should be completed first. Thus, $(3 - 1) \div 2$ means one should first subtract 1 from 3, and then divide that result by 2.

The **order of operations** gives an order for how a mathematical expression is to be simplified:

- Parentheses
- Exponents
- Multiplication
- Division
- Addition
- Subtraction

To help remember this, many students like to use the mnemonic PEMDAS. Some students associate this word with a phrase to help them, such as "Pirates Eat Many Donuts at Sea." Here is a quick example:

Evaluate $2^2 \times (3 - 1) \div 2 + 3$.

Parenthesis: $2^2 \times 2 \div 2 + 3$.

Exponents: $4 \times 2 \div 2 + 3$

Multiply: $8 \div 2 + 3$.

Divide: $4 + 3$.

Addition: 7

In solving multi-step problems, the first step is to line up the available information. Then, try to decide what information the problem is asking to be found. Once this is determined, construct a strip diagram to display the known information along with any information to be calculated. Finally, the missing information can be represented by a **variable** (a letter from the alphabet that represents a number) in a mathematical equation that the student can solve.

For example, Delilah collects stickers and her friends gave her some stickers to add to her current collection. Joe gave her 45 stickers, and Aimee gave her 2 times the number of stickers that Joe gave Delilah. How many stickers did Delilah have to start with, if after her friends gave her more stickers, she had a total of 187 stickers?

In order to solve this, the given information must first be sorted out. Joe gives Delilah 45 stickers, Aimee gives Delilah 2 times the number Joe gives (2 × 45), and the end total of stickers is 187.

A strip diagram represents these numbers as follows:

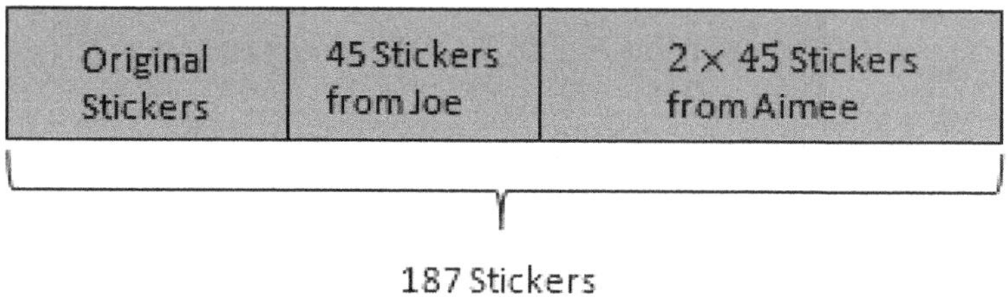

The entire situation can be modeled by this equation, using the variable s to stand for the original number of stickers:

$$s + 45 + (2 \times 45) = 187.$$

Solving for s would give the solution, as follows:

$$s + 45 + 90 = 187$$

$$s + 135 = 187$$

$$s + 135 - 135 = 187 - 135$$

$$s = 52 \text{ stickers.}$$

The Reasonableness of Answers

When solving math word problems, the solution obtained should make sense within the given scenario. The step of checking the solution will reduce the possibility of a calculation error or a solution that may be *mathematically* correct but not applicable in the real world. Consider the following scenarios:

A problem states that Lisa got 24 out of 32 questions correct on a test and asks to find the percentage of correct answers. To solve the problem, a student divided 32 by 24 to get 1.33, and then multiplied by 100 to get 133 percent. By examining the solution within the context of the problem, the student should recognize that getting all 32 questions correct will produce a perfect score of 100 percent. Therefore, a score of 133 percent with 8 incorrect answers does not make sense and the calculations should be checked.

A problem states that the maximum weight on a bridge cannot exceed 22,000 pounds. The problem asks to find the maximum number of cars that can be on the bridge at one time if each car weighs 4,000 pounds. To solve this problem, a student divided 22,000 by 4,000 to get an answer of 5.5. By examining the solution within the context of the problem, the student should recognize that although the calculations are mathematically correct, the solution does not make sense. Half of a car on a bridge is not possible, so the student should determine that a maximum of 5 cars can be on the bridge at the same time.

Estimating

Estimation is finding a value that is close to a solution but is not the exact answer. For example, if there are values in the thousands to be multiplied, then each value can be estimated to the nearest thousand and the calculation performed. This value provides an approximate solution that can be determined very quickly.

Rounding is the process of either bumping a number up or down, based on a specified place value. First, the place value is specified. Then, the digit to its right is looked at. For example, if rounding to the nearest hundreds place, the digit in the tens place is used. If it is a 0, 1, 2, 3, or 4, the digit being rounded to is left alone. If it is a 5, 6, 7, 8 or 9, the digit being rounded to is increased by one. All other digits before the decimal point are then changed to zeros, and the digits in decimal places are dropped. If a decimal place is being rounded to, all subsequent digits are just dropped. For example, if 845,231.45 was to be rounded to the nearest thousands place, the answer would be 845,000. The 5 would remain the same due to the 2 in the hundreds place. Also, if 4.567 was to be rounded to the nearest tenths place, the answer would be 4.6. The 5 increased to 6 due to the 6 in the hundredths place, and the rest of the decimal is dropped.

Sometimes when performing operations such as multiplying numbers, the result can be estimated by rounding. For example, to estimate the value of 11.2×2.01, each number can be rounded to the nearest integer. This will yield a result of 22.

Rounding numbers helps with estimation because it changes the given number to a simpler, although less accurate, number than the exact given number. Rounding allows for easier calculations, which estimate the results of using the exact given number. The accuracy of the estimate and ease of use depends on the place value to which the number is rounded. Rounding numbers consists of:

- determining what place value the number is being rounded to
- examining the digit to the right of the desired place value to decide whether to round up or keep the digit, and
- replacing all digits to the right of the desired place value with zeros.

To round 746,311 to the nearest ten thousand, the digit in the ten thousands place should be located first. In this case, this digit is 4 (7<u>4</u>6,311). Then, the digit to its right is examined. If this digit is 5 or greater, the number will be rounded up by increasing the digit in the desired place by one. If the digit to the right of the place value being rounded is 4 or less, the number will be kept the same. For the given example, the digit being examined is a 6, which means that the number will be rounded up by increasing the digit to the left by one. Therefore, the digit 4 is changed to a 5. Finally, to write the rounded number, any digits to the left of the place value being rounded remain the same and any to its right are replaced with zeros. For the given example, rounding 746,311 to the nearest ten thousand will produce 750,000. To round 746,311 to the nearest hundred, the digit to the right of the three in the hundreds place is examined to determine whether to round up or keep the same number. In this case, that digit is a 1, so the number will be kept the same and any digits to its right will be replaced with zeros. The resulting rounded number is 746,300.

Rounding place values to the right of the decimal follows the same procedure, but digits being replaced by zeros can simply be dropped. To round 3.752891 to the nearest thousandth, the desired place value is located (3.75<u>2</u>891) and the digit to the right is examined. In this case, the digit 8 indicates that the number will be rounded up, and the 2 in the thousandths place will increase to a 3. Rounding up and replacing the digits to the right of the thousandths place produces 3.753000 which is equivalent to 3.753. Therefore, the zeros are not necessary and the rounded number should be written as 3.753.

When rounding up, if the digit to be increased is a 9, the digit to its left is increased by 1 and the digit in the desired place value is changed to a zero. For example, the number 1,598 rounded to the nearest ten is 1,600. Another example shows the number 43.72961 rounded to the nearest thousandth is 43.730 or 43.73.

Mental math should always be considered as problems are worked through, and the ability to work through problems in one's head helps save time. If a problem is simple enough, such as $15 + 3 = 18$, it should be completed mentally. The ability to do this will increase once addition and subtraction in higher place values are grasped. Also, mental math is important in multiplication and division. The times tables multiplying all numbers from 1 to 12 should be memorized. This will allow for division within those numbers to be memorized as well. For example, we should know easily that $121 \div 11 = 11$ because it should be memorized that $11 \times 11 = 121$.

Here is the multiplication table to be memorized:

x	1	2	3	4	5	6	7	8	9	10	11	12	13	14	15
1	1	2	3	4	5	6	7	8	9	10	11	12	13	14	15
2	2	4	6	8	10	12	14	16	18	20	22	24	26	28	30
3	3	6	9	12	15	18	21	24	27	30	33	36	39	42	45
4	4	8	12	16	20	24	28	32	36	40	44	48	52	56	60
5	5	10	15	20	25	30	35	40	45	50	55	60	65	70	75
6	6	12	18	24	30	36	42	48	54	60	66	72	78	84	90
7	7	14	21	28	35	42	49	56	63	70	77	84	91	98	105
8	8	16	24	32	40	48	56	64	72	80	88	96	104	112	120
9	9	18	27	36	45	54	63	72	81	90	99	108	117	126	135
10	10	20	30	40	50	60	70	80	90	100	110	120	130	140	150
11	11	22	33	44	55	66	77	88	99	110	121	132	143	154	165
12	12	24	36	48	60	72	84	96	108	120	132	144	156	168	180
13	13	26	39	52	65	78	91	104	117	130	143	156	169	182	195
14	14	28	42	56	70	84	98	112	126	140	154	168	182	196	210
15	15	30	45	60	75	90	105	120	135	150	165	180	195	210	225

The values along the diagonal of the table consist of **perfect squares**. A perfect square is a number that represents a product of two equal integers.

Estimating Absolute and Relative Error in the Numerical Answer to a Problem

Once a result is determined to be logical within the context of a given problem, the result should be evaluated by its nearness to the expected answer. This is performed by approximating given values to perform mental math. Numbers should be rounded to the nearest value possible to check the initial results.

Consider the following example: A problem states that a customer is buying a new sound system for their home. The customer purchases a stereo for \$435, 2 speakers for \$67 each, and the necessary cables for \$12. The customer chooses an option that allows him to spread the costs over equal payments for 4 months. How much will the monthly payments be?

After making calculations for the problem, a student determines that the monthly payment will be \$145.25. To check the accuracy of the results, the student rounds each cost to the nearest ten ($440 + 70 + 70 + 10$) and determines that the total is approximately \$590. Dividing by 4 months gives an approximate monthly payment of \$147.50. Therefore, the student can conclude that the solution of \$145.25 is very close to what should be expected.

When rounding, the place-value that is used in rounding can make a difference. Suppose the student had rounded to the nearest hundred for the estimation. The result ($400 + 100 + 100 + 0 = 600;\ 600 \div 4 = 150$) will show that the answer is reasonable, but not as close to the actual value as rounding to the nearest ten.

When considering the accuracy of estimates, the error in the estimated solution can be shown as absolute and relative. **Absolute error** tells the actual difference between the estimated value and the true, calculated value. The **relative error** tells how large the error is in relation to the true value. There may be two problems where the absolute error of the values (the estimated one and the calculated one) is 10. For one problem, this may mean the relative error in the estimate is very small because the estimated value is 15,000, and the true value is 14,990. Ten in relation to the true value of 15,000 is small: 0.06%. For the other problem, the estimated value is 50, and the true value is 40. In this case, the absolute error of 10 means a high relative error because the true value is smaller. The relative error is 10/40 = 0.25 or 25%.

Using Variables to Represent Quantities and Constructing Simple Equations and Inequalities to Solve Problems by Reasoning About the Quantities

Solving Word Problems Leading to Equations of the Form $px + q = r$ and $p(x + q) = r$

Linear expressions and equations are concise mathematical statements that can be written to model a variety of scenarios. Questions found pertaining to this topic will contain one variable only. A variable is an unknown quantity, usually denoted by a letter (*x, n, p,* etc.). In the case of linear expressions and equations, the power of the variable (its exponent) is 1. A variable without a visible exponent is raised to the first power.

A linear expression is a statement about an unknown quantity expressed in mathematical symbols. The statement "five times a number added to forty" can be expressed as $5x + 40$. A linear equation is a statement in which two expressions (at least one containing a variable) are equal to each other. The statement "five times a number added to forty is equal to ten" can be expressed as $5x + 40 = 10$.

Real world scenarios can also be expressed mathematically. Suppose a job pays its employees $300 per week and $40 for each sale made. The weekly pay is represented by the expression $40x + 300$ where *x* is the number of sales made during the week.

Consider the following scenario: Bob had $20 and Tom had $4. After selling 4 ice cream cones to Bob, Tom has as much money as Bob. The cost of an ice cream cone is an unknown quantity and can be represented by a variable (*x*). The amount of money Bob has after his purchase is four times the cost of an ice cream cone subtracted from his original $\$20 \rightarrow 20 - 4x$. The amount of money Tom has after his sale is four times the cost of an ice cream cone added to his original $\$4 \rightarrow 4x + 4$. After the sale, the amount of money that Bob and Tom have are equal $\rightarrow 20 - 4x = 4x + 4$. Solving for x yields $x = 2$.

When expressing a verbal or written statement mathematically, it is key to understand words or phrases that can be represented with symbols. The following are examples:

Symbol	**Phrase**
$+$	added to, increased by, sum of, more than
$-$	decreased by, difference between, less than, take away
x	multiplied by, 3 (4, 5 . . .) times as large, product of
$\div$	divided by, quotient of, half (third, etc.) of
$=$	is, the same as, results in, as much as
$x, t, n, etc.$	a number, unknown quantity, value of

When asked to solve a linear equation, one must determine a numerical value for the unknown variable. Given a linear equation involving addition, subtraction, multiplication, and division, isolation of the variable is done by working backward. Addition and subtraction are inverse operations, as are multiplication and division; therefore, they can be used to cancel each other out.

The first steps to solving linear equations are to distribute if necessary and combine any like terms that are on the same side of the equation. Sides of an equation are separated by an $=$ sign. Next, the equation should be manipulated to get the variable on one side. Whatever is done to one side of an equation, must be done to the other side to remain equal. Then, the variable should be isolated by using inverse operations to undo the order of operations backward. Undo addition and subtraction, then undo multiplication and division. For example:

Solve $4(t-2)+2t-4=2(9-2t)$

Distribute: $4t-8+2t-4=18-4t$

Combine like terms: $6t-12=18-4t$

Add 4t to each side to move the variable: $10t-12=18$

Add 12 to each side to isolate the variable: $10t=30$

Divide each side by 10 to isolate the variable: $t=3$

The answer can be checked by substituting the value for the variable into the original equation and ensuring both sides calculate to be equal.

In word problems, multiple quantities are often provided with a request to find some kind of relation between them. This often will mean that one variable (the dependent variable whose value needs to be found) can be written as a function of another variable (the independent variable whose value can be figured from the given information). The usual procedure for solving these problems is to start by giving each quantity in the problem a variable, and then figuring the relationship between these variables.

For example, suppose a car gets 25 miles per gallon. How far will the car travel if it uses 2.4 gallons of fuel? In this case, *y* would be the distance the car has traveled in miles, and *x* would be the amount of fuel burned in gallons (2.4). Then the relationship between these variables can be written as an algebraic equation, $y=25x$. In this case, the equation is $y=25\times 2.4=60$, so the car has traveled 60 miles.

Some word problems require more than just one simple equation to be written and solved. Consider the following situations and the linear equations used to model them.

Suppose Margaret is 2 miles to the east of John at noon. Margaret walks to the east at 3 miles per hour. How far apart will they be at 3 p.m.? To solve this, *x* would represent the time in hours past noon, and *y* would represent the distance between Margaret and John. Now, noon corresponds to the equation where *x* is 0, so the *y* intercept is going to be 2. It's also known that the slope will be the rate at which the distance is changing, which is 3 miles per hour. This means that the slope will be 3 (be careful at this point: if units were used, other than miles and hours, for *x* and *y* variables, a conversion of the given information to the appropriate units would be required first). The simplest way to write an equation given the *y*-intercept, and the slope is the Slope-Intercept form, is $y=mx+b$. Recall that m here is the slope and *b* is the *y* intercept. So, $m=3$ and $b=2$. Therefore, the equation will be $y=3x+2$. The word problem asks how far to the east Margaret will be from John at 3 p.m., which means when *x* is 3. So, substitute $x=3$ into this equation to obtain:

$$y=3\times 3+2=9+2=11$$

Therefore, she will be 11 miles to the east of him at 3 p.m.

For another example, suppose that a box with 4 cans in it weighs 6 lbs., while a box with 8 cans in it weighs 12 lbs. Find out how much a single can weighs. To do this, let *x* denote the number of cans in the box, and *y* denote the weight of the box with the cans in lbs. This line touches two pairs: $(4, 6)$ and (8, 12). A formula for this relation could be written using the two-point form, with $x_1 = 4, y_1 = 6, x_2 = 8, y_2 = 12$. This would yield $\frac{y-6}{x-4} = \frac{12-6}{8-4}$, or $\frac{y-6}{x-4} = \frac{6}{4} = \frac{3}{2}$. However, only the slope is needed to solve this problem, since the slope will be the weight of a single can. From the computation, the slope is $\frac{3}{2}$. Therefore, each can weighs $\frac{3}{2}$ lb.

The following problem is an example of using rational expressions:

Reggie wants to lay sod in his rectangular backyard. The length of the yard is given by the expression $4x + 2$ and the width is unknown. The area of the yard is $20x + 10$. Reggie needs to find the width of the yard. Knowing that the area of a rectangle is length multiplied by width, an expression can be written to find the width: $\frac{20x+10}{4x+2}$, area divided by length. Simplifying this expression by factoring out 10 on the top and 2 on the bottom leads to this expression:

$$\frac{10(2x+1)}{2(2x+1)}$$

Canceling out the $2x + 1$ results in $\frac{10}{2} = 5$. The width of the yard is found to be 5 by simplifying the rational expression.

<u>Solving Word Problems Leading to Inequalities of the Form $px + q > r$ or $px + q < r$</u>

Linear inequalities are a concise mathematical way to express the relationship between unequal values. More specifically, they describe in what way the values are unequal. A value could be greater than (>); less than (<); greater than or equal to (≥); or less than or equal to (≤) another value. The statement "five times a number added to forty is more than sixty-five" can be expressed as $5x + 40 > 65$. Common words and phrases that express inequalities are:

Symbol	Phrase
$<$	is under, is below, smaller than, beneath
$>$	is above, is over, bigger than, exceeds
$\leq$	no more than, at most, maximum
$\geq$	no less than, at least, minimum

When solving a linear inequality, the solution is the set of all numbers that makes the statement true. The inequality $x + 2 \geq 6$ has a solution set of 4 and every number greater than 4 (4.0001, 5, 12, 107, etc.). Adding 2 to 4 or any number greater than 4 would result in a value that is greater than or equal to 6. Therefore, $x \geq 4$ would be the solution set.

Solution sets for linear inequalities often will be displayed using a number line. If a value is included in the set ($\geq$ or $\leq$), there is a shaded dot placed on that value and an arrow extending in the direction of the solutions. For a variable $>$ or $\geq$ a number, the arrow would point right on the number line (the direction where the numbers increase); and if a variable is $<$ or $\leq$ a number, the arrow would point left (where the numbers decrease). If the value is not included in the set ($>$ or $<$), an open circle on that value would be used with an arrow in the appropriate direction.

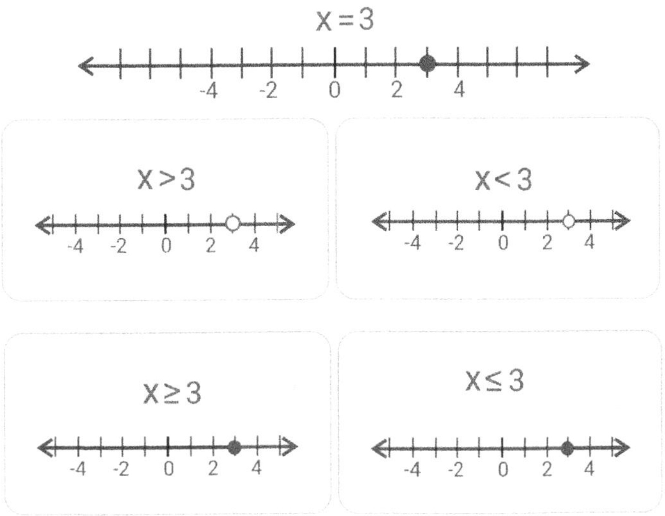

Students may be asked to write a linear inequality given a graph of its solution set. To do so, they should identify whether the value is included (shaded dot or open circle) and the direction in which the arrow is pointing.

In order to algebraically solve a linear inequality, the same steps should be followed as in solving a linear equation. The inequality symbol stays the same for all operations EXCEPT when dividing by a negative number. If dividing by a negative number while solving an inequality, the relationship reverses (the sign flips). Dividing by a positive does not change the relationship, so the sign stays the same. In other words, $>$ switches to $<$ and vice versa. An example is shown below.

Solve $-2(x+4) \leq 22$

Distribute: $-2x - 8 \leq 22$

Add 8 to both sides: $-2x \leq 30$

Divide both sides by -2: $x \geq 15$

Geometry

Drawing, Constructing, and Describing Geometrical Figures and Describing the Relationships Between Them

Solving Problems Involving Scale Drawings of Geometric Figures

Scale drawings are used in designs to model the actual measurements of a real-world object. For example, the blueprint of a house might indicate that it is drawn at a scale of 3 inches to 8 feet. Given one value and asked to determine the width of the house, a proportion should be set up to solve the problem. Given the scale of 3in:8ft and a blueprint width of 1 ft (12 in.), to find the actual width of the building, the proportion $\frac{3}{8} = \frac{12}{x}$ should be used. This results in an actual width of 32 ft.

Drawing Geometric Shapes With Given Conditions

Drawing geometric shapes involves knowing the basic properties of the shapes, such as the number of sides and angles. Drawing itself may be freehand, with ruler and protractor, or with technology. The tools needed to make formal geometric constructions are a compass, a ruler, paper folding, or geometry software. These tools can be used to copy or bisect (split into two equal halves) a line segment or angle. They can also replicate a specific shape and construct perpendicular lines, a perpendicular bisector (a line that bisects) of a line segment, and a line parallel to a given line through a specified point.

Beginning with formal geometric constructions, various geometric figures and shapes can be built. Definitions and theorems for lines and angles can be used in parallel with geometric constructions to build shapes such as equilateral triangles, squares, and rectangular hexagons inscribed in a circle. Definitions of shapes involving congruence of sides and angles within each type of figure must be understood and used in parallel with constructing congruent line segments, parallel and perpendicular lines, and congruent angles.

Diagonals are lines (excluding sides) that connect two vertices within a polygon. **Mutually bisecting diagonals** intersect at their midpoints. Parallelograms, rectangles, squares, and rhombuses have mutually bisecting diagonals. However, trapezoids don't have such lines. **Perpendicular diagonals** occur when they form four right triangles at their point of intersection. Squares and rhombuses have perpendicular diagonals, but trapezoids, rectangles, and parallelograms do not. Finally, **perpendicular bisecting** diagonals (also known as **perpendicular bisectors**) form four right triangles at their point of intersection, but this intersection is also the midpoint of the two lines. Both rhombuses and squares have perpendicular bisecting angles, but trapezoids, rectangles, and parallelograms do not. Knowing these definitions can help tremendously in problems that involve both angles and diagonals.

A **polygon** is a closed geometric figure in a plane (flat surface) consisting of at least 3 sides formed by line segments.

Polygons can be either convex or concave. A polygon that has interior angles all measuring less than 180° is **convex**. A **concave** polygon has one or more interior angles measuring greater than 180°. Examples are shown below.

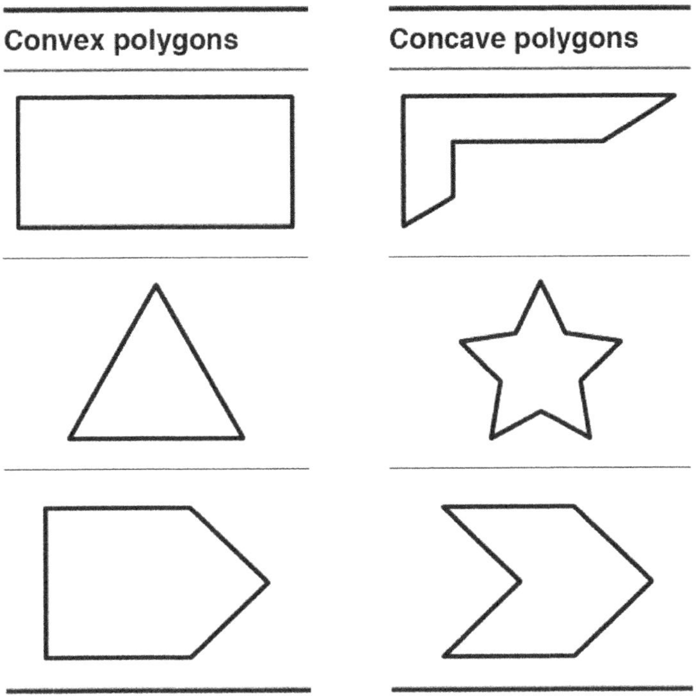

Polygons are often defined as two-dimensional shapes. Common two-dimensional shapes include circles, triangles, squares, rectangles, pentagons, and hexagons. A **circle** is a two-dimensional shape without sides, so it is not a polygon.

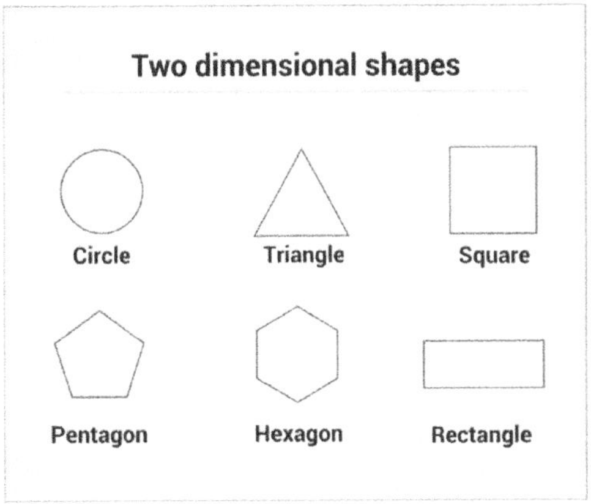

Polygons can be classified by the number of sides (also equal to the number of angles) they have. The following are the names of polygons with a given number of sides or angles:

# of Sides	Name of Polygon
3	Triangle
4	Quadrilateral
5	Pentagon
6	Hexagon
7	Septagon (or heptagon)
8	Octagon
9	Nonagon
10	Decagon

Equiangular polygons are polygons in which the measure of every interior angle is the same. The sides of equilateral polygons are always the same length. If a polygon is both equiangular and equilateral, the polygon is defined as a **regular polygon**. Examples are shown below:

Many four-sided figures can also be identified using properties of angles and lines. A **quadrilateral** is a closed shape with four sides. A **parallelogram** is a specific type of quadrilateral that has two sets of parallel lines that have the same length. A **trapezoid** is a quadrilateral having only one set of parallel sides. A **rectangle** is a parallelogram that has four right angles. A **rhombus** is a parallelogram with two acute angles, two obtuse angles, and four equal sides. The acute angles are of equal measure, and the obtuse angles are of equal measure. Finally, a **square** is a rhombus consisting of four right angles. It is important to note that some of these shapes share common attributes. For instance, all four-sided shapes are quadrilaterals.

All squares are rectangles, but not all rectangles are squares.

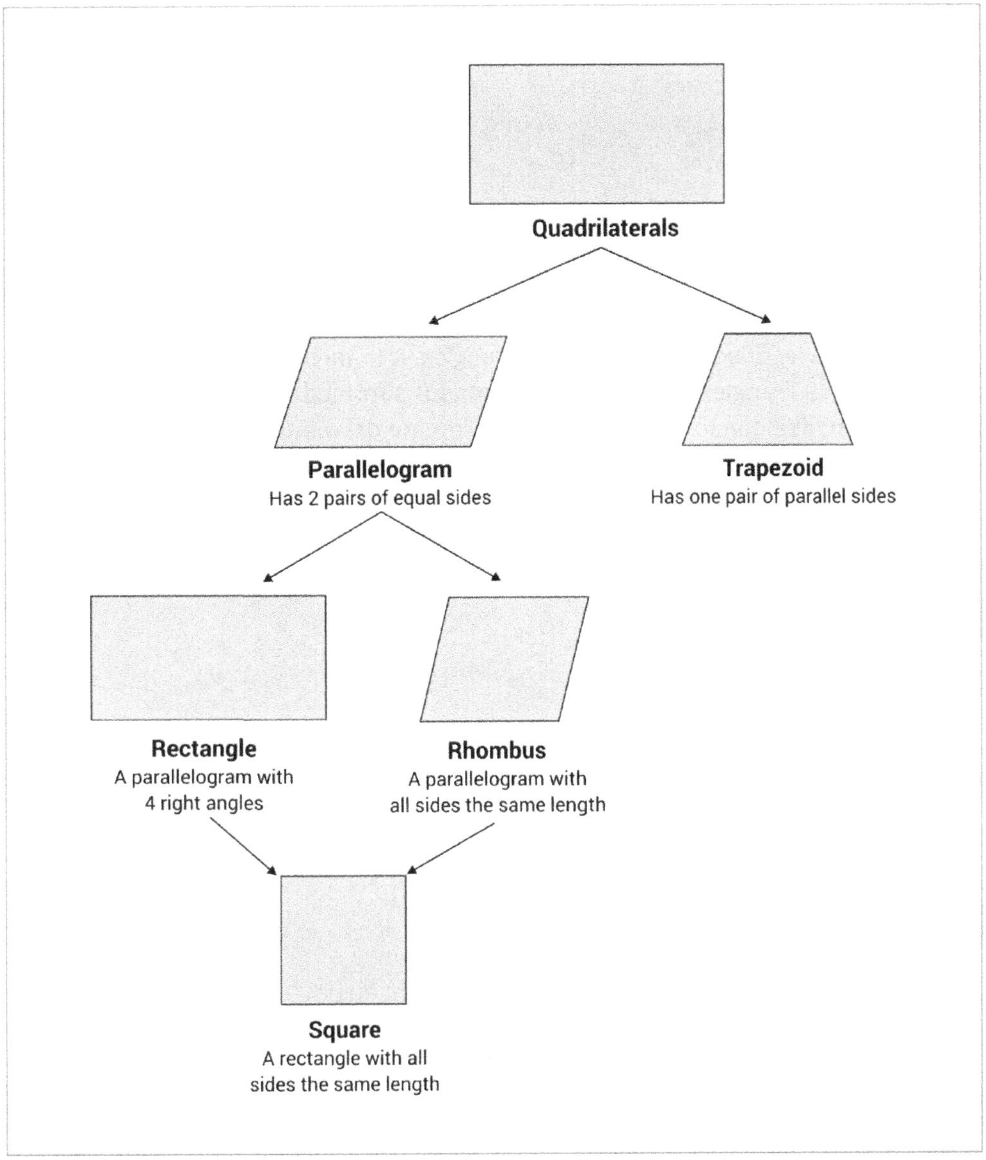

Consider the following:

Which of the following are not rectangles?

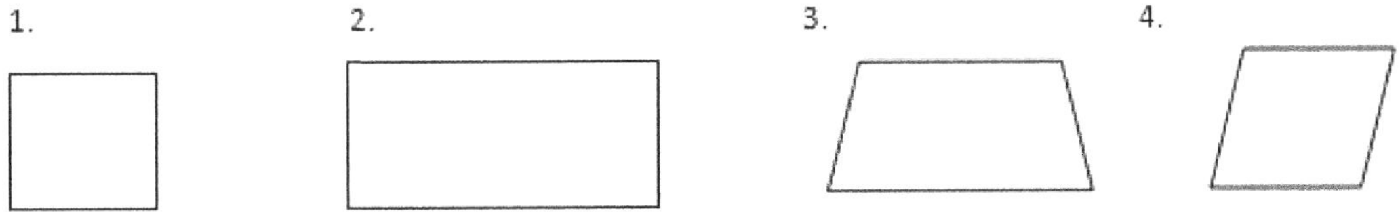

Even though all have four sides and some are even parallel sides, numbers 3 and 4 do not fit the definition of a rectangle.

Which of the following are not parallelograms?

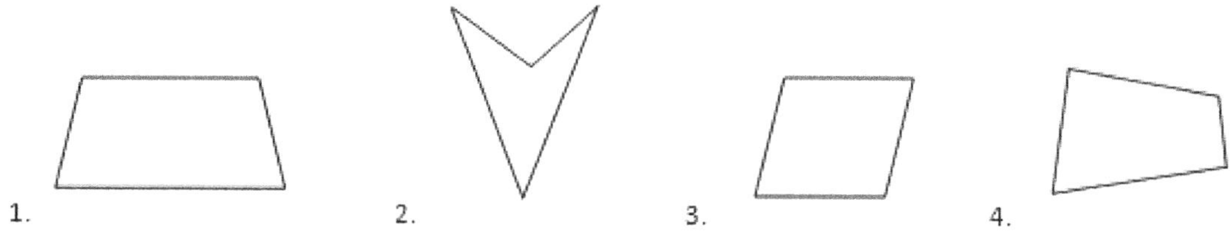

Even though all have four sides, numbers 2 and 4 do not fit the definition of a parallelogram because they have no parallel sides.

Triangles can be further classified by their sides and angles. A triangle with its largest angle measuring 90° is a **right triangle**. A 90° angle, also called a **right angle,** is formed from two perpendicular lines. It looks like a hard corner, like that in a square. The little square draw into the angle is the symbol used to denote that that angle is indeed a right angle. Any time that symbol is used, it denotes the measure of the angle is 90°. Below is a picture of a right angle, and below that, a right triangle.

A right angle:

Here is a right triangle, which is a triangle that contains a right angle:

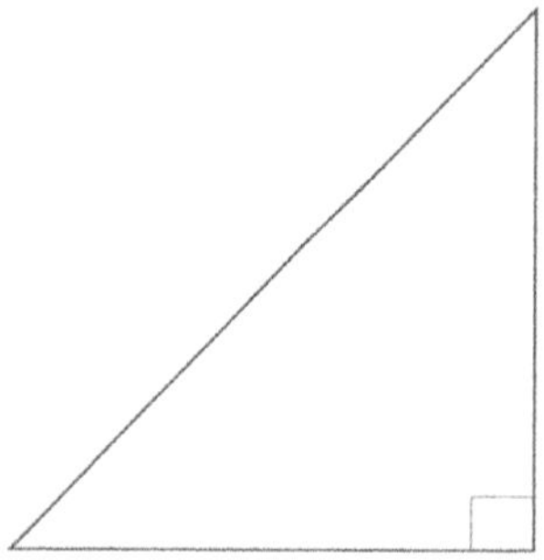

A triangle with the largest angle less than 90° is an acute triangle. A triangle with the largest angle greater than 90° is an obtuse triangle.

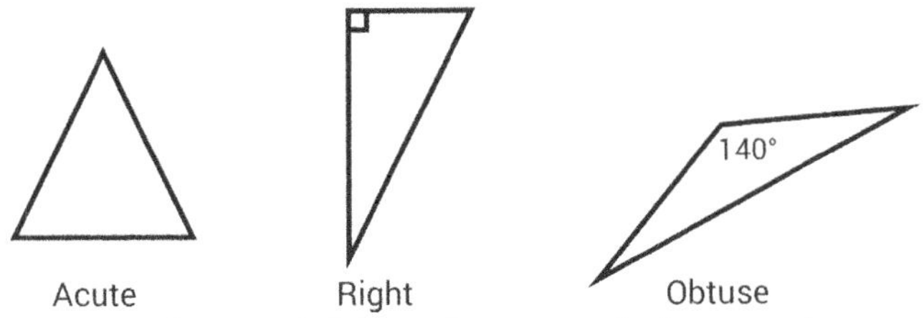

Acute
3 acute angles

Right
1 right angle

Obtuse
1 obtuse angle

A triangle consisting of two equal sides and two equal angles is an isosceles triangle. A triangle with three equal sides and three equal angles is an equilateral triangle. A triangle with no equal sides or angles is a scalene triangle.

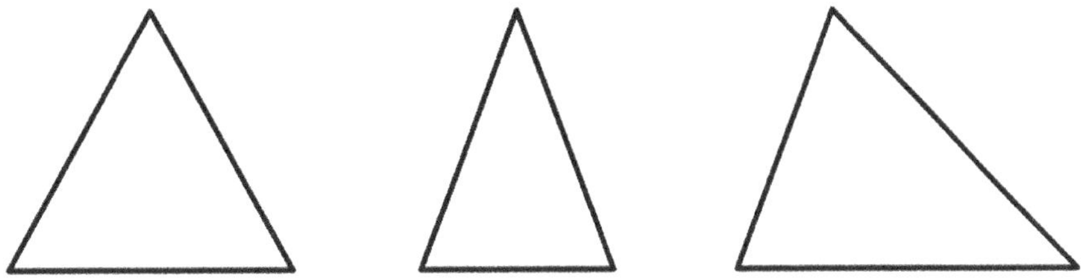

Equilateral Triangle **Isosceles Triangle** **Scalene Triangle**

Mathematical problems involving polygons with more than four sides usually involve side length and angle measurements. The sum of all internal angles in a polygon equals $180(n-2)$ degrees, where n is the number of sides. Therefore, the total of all internal angles in a pentagon is 540 degrees because there are five sides so $180(5-2)$ = 540 degrees. Unfortunately, area formulas don't exist for polygons with more than four sides. However, their shapes can be split up into triangles, and the formula for area of a triangle can be applied and totaled to obtain the area for the entire figure.

Protractors are used to draw angles with a given measure. The little hole at the midpoint along the bottom flat diameter of the circle is where the dot for the vertex of the angle should be drawn. Then a straight line is drawn using the straight edge of the protractor. This will be the ray of the angle that runs through the 0^0 mark on the protractor (again, along the straight part along the bottom). Then, look around the outer rim of the protractor until you find the desired number for the angle you want to draw. For example, if tasked with drawing a 40^0 angle, look for the number 40. Then draw a small point along the outer edge of the protractor where the 40^0 mark is (or whatever mark you need for your specific angle). Then, use the straight edge to draw a ray that runs from the vertex through the point you just made. You now should have drawn the appropriate angle.

Describing the Two-Dimensional Figures That Result From Slicing Three-Dimensional Figures

Two-dimensional objects are formed when three-dimensional objects are "sliced" in various ways. For example, any cross section of a sphere is a circle. Some three-dimensional objects have different cross sections depending on how the object is sliced. For example, the cross section of a cylinder can be a circle or a rectangle, and the cross section of a pyramid can be a square or a triangle. In addition, three-dimensional objects can be formed by rotating two-dimensional objects. Certain rotations can relate the two-dimensional cross sections back to the original three-dimensional objects. The objects must be rotated around an imaginary line known as the **rotation axis**. For example, a right triangle can be rotated around one of its legs to form a cone. A sphere can be formed by rotating a semicircle around a line segment formed from its diameter. Finally, rotating a square around one of its sides forms a cylinder.

Three-dimensional objects can be simplified into related two-dimensional shapes to solve problems. This simplification can make problem-solving a much easier experience. An isometric representation of a three-dimensional object can be completed so that important properties (e.g., shape, relationships of faces and surfaces) are noted. Edges and vertices can be translated into two-dimensional objects as well. For example, below is a three-dimensional object that's been partitioned into two-dimensional representations of its faces:

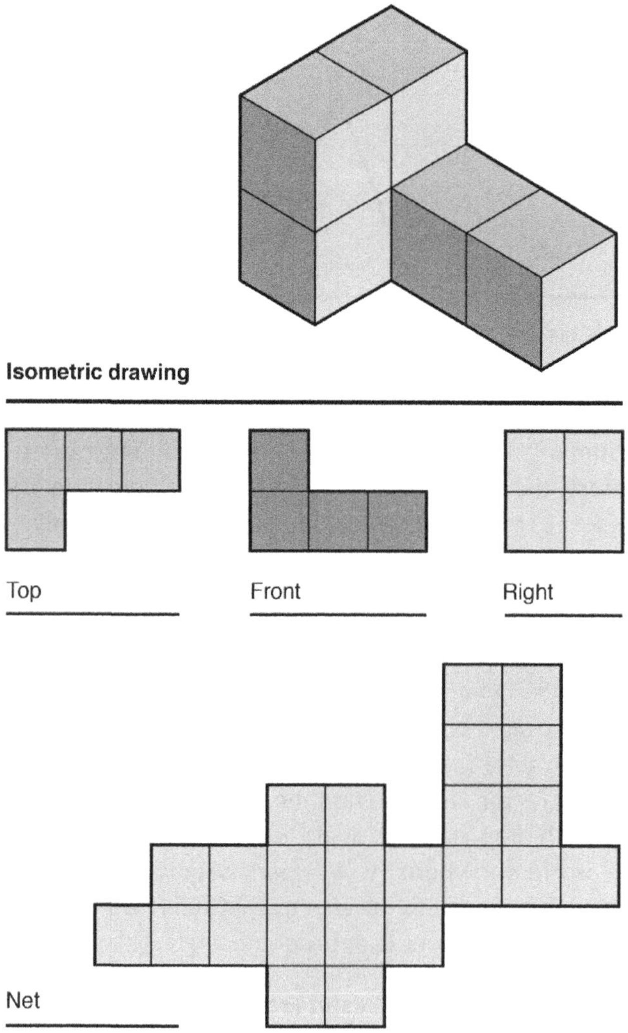

The net represents the sum of the three different faces. Depending on the problem, using a smaller portion of the given shape may be helpful, by simplifying the steps necessary to solve.

Solving Real-Life and Mathematical Problems Involving Angle Measure, Area, Surface Area, and Volume

Using Formulas for the Area and Circumference of a Circle to Solve Problems

A circle's perimeter—also known as its **circumference**—is measured by multiplying the diameter by π.

Diameter is the straight line measured from one end to the direct opposite end of the circle.

π is referred to as pi and is equal to 3.14 (with rounding).

So the formula is $\pi \times d$.

This is sometimes expressed by the formula $C = 2 \times \pi \times r$, where r is the radius of the circle. These formulas are equivalent, as the radius equals half of the diameter.

As mentioned, the **diameter** of a circle is the straight line measured from one end to the direct opposite end of the circle, while the radius is half that distance. Essentially, the **radius** of a circle is the distance from the center of the circle to any point on the circle. For example, if the diameter of a circle is 10 inches, the radius is 5 inches.

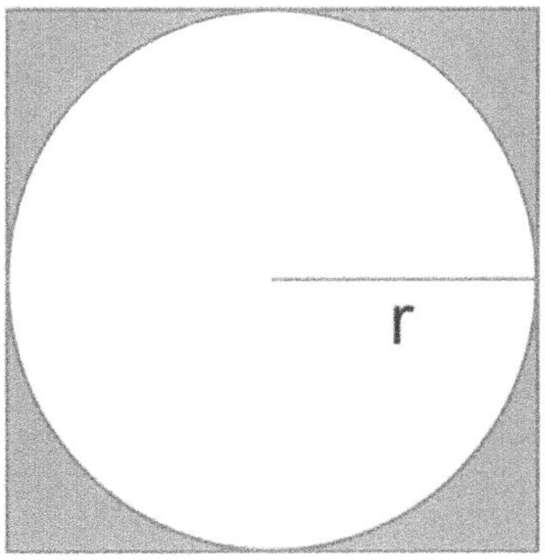

The area of a circle is calculated through the formula $A = \pi \times r^2$. For example, if a circle has a radius of 5 centimeters, the area is computed by substituting 5 for the radius: $\pi \times (5)^2 = 25\pi\ sq.cm$. The test will indicate either to leave the answer with π attached or to calculate to the nearest decimal place, which means multiplying by 3.14 for π. For the previous example, this would be $25 \times 3.14 = 78.5\ sq.cm$.

Consider the following problem: A pizzeria owner regularly creates jumbo pizzas, each with a radius of 9 inches. She is mathematically inclined, and wants to know the area of the pizza to purchase the correct boxes and know how much she is feeding her customers. What is the area of the circle, in terms of π, with a radius of 9 inches?

To solve this problem, the formula for the area of the circle (πr^2). The radius of the pizza is 9 inches so 9 squared is 81. Therefore, the area of the pizza will be 81 π in^2.

Using Facts About Supplementary, Complementary, Vertical, and Adjacent Angles in a Multi-Step Problem to Write and Solve Equations for an Unknown Angle

An **angle** consists of two rays that have a common endpoint. This common endpoint is called the **vertex** of the angle. The two rays can be called sides of the angle. The angle below has a vertex at point *B* and the sides consist of ray *BA* and ray *BC*. An angle can be named in three ways:

1. Using the vertex and a point from each side, with the vertex letter in the middle.
2. Using only the vertex. This can only be used if it is the only angle with that vertex.
3. Using a number that is written inside the angle.

The angle below can be written $\angle ABC$ (read angle *ABC*), $\angle CBA$, $\angle B$, or $\angle 1$.

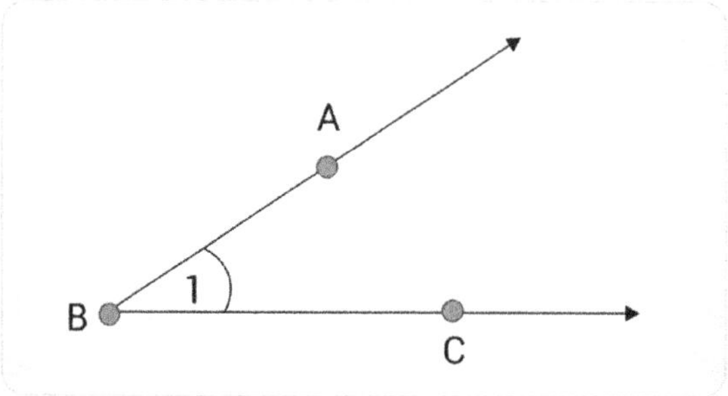

An angle divides a **plane**, or flat surface, into three parts: the angle itself, the interior (inside) of the angle, and the exterior (outside) of the angle. The figure below shows point *M* on the interior of the angle and point *N* on the exterior of the angle.

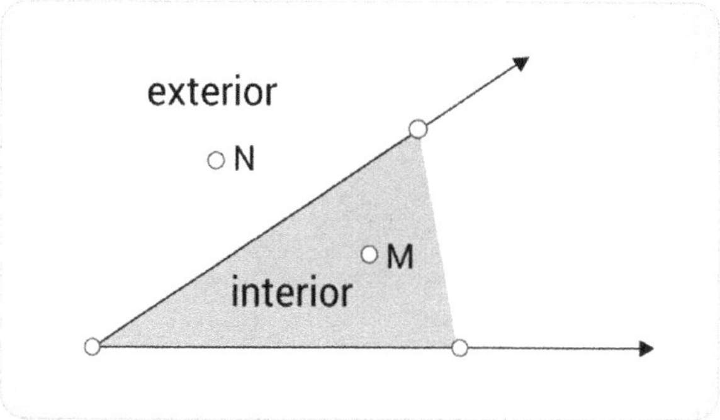

Angles can be measured in units called degrees, with the symbol °. The degree measure of an angle is between 0° and 180°, is a measure of rotation, and can be obtained by using a **protractor**, which is the tool pictured below.

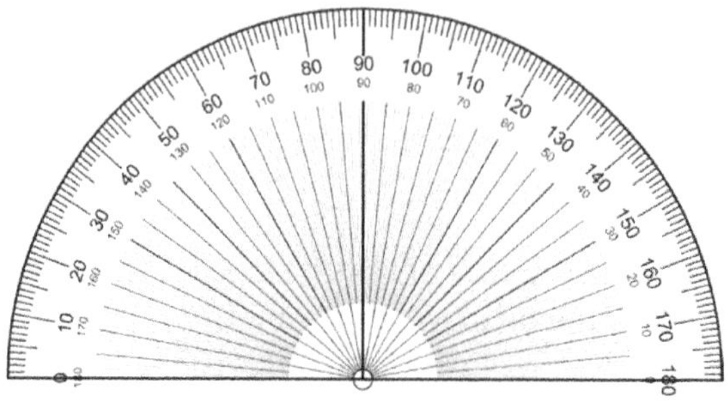

To use a protractor to measure an angle, the vertex, or corner, of the angle goes in the midpoint of the protractor, in that small circle along the bottom straight edge. Then one line of the angle is lined up along the bottom edge, toward the 0° indicator. Then, the degrees are read wherever the other line of the angle crosses.

In the example below, it can be seen that the angle measures about 30°:

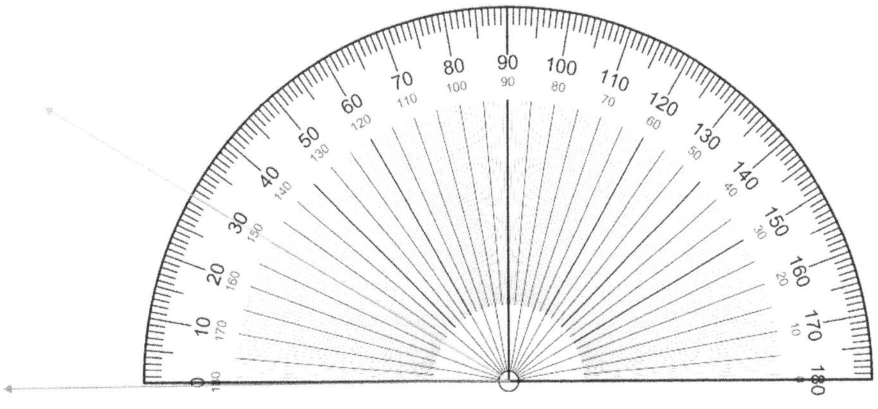

A straight angle (or simply a line) measures exactly 180°. A right angle's sides meet at the vertex to create a square corner. A right angle measures exactly 90° and is typically indicated by a box drawn in the interior of the angle. An acute angle has an interior that is narrower than a right angle. The measure of an acute angle is any value less than 90° and greater than 0°. For example, 89.9°, 47°, 12°, and 1°. An obtuse angle has an interior that is wider than a right angle. The measure of an obtuse angle is any value greater than 90° but less than 180°. For example, 90.1°, 110°, 150°, and 179.9°. Any two angles that sum up to 90 degrees are known as **complementary angles**.

- Acute angles: Less than 90°
- Obtuse angles: Greater than 90°
- Right angles: 90°
- Straight angles: 180°

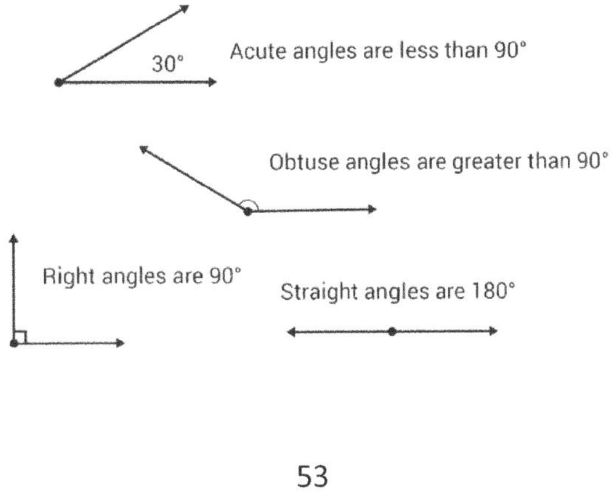

Supplementary angles add up to 180 degrees. **Vertical angles** are two nonadjacent angles formed by two intersecting lines. For example, in the following picture, angles 4 and 2 are vertical angles and so are angles 1 and 3:

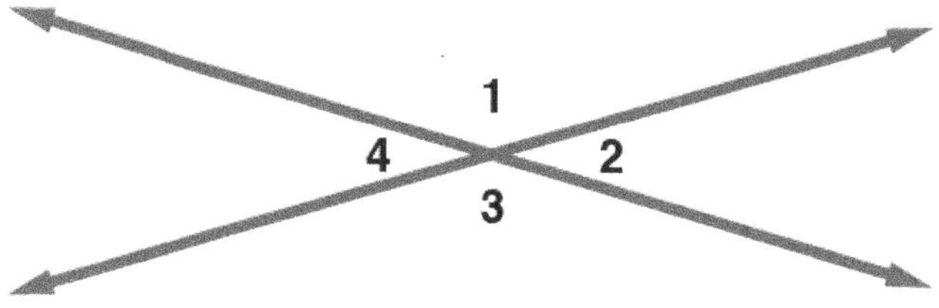

Vertical Angles

Corresponding angles are two angles in the same position whenever a straight line (known as a **transversal**) crosses two others. If the two lines are parallel, the corresponding angles are equal. In the following diagram, angles 1 and 3 are corresponding angles but aren't equal to each other:

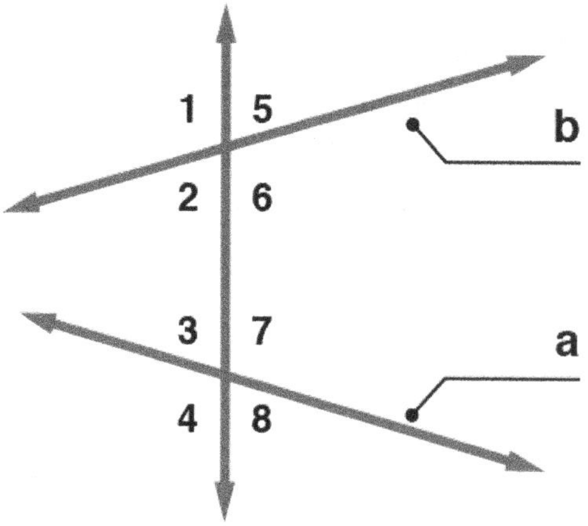

Corresponding Angles

Alternate interior angles are also a pair of angles formed when two lines are crossed by a transversal. They are opposite angles that exist inside of the two lines. In the corresponding angles diagram above, angles 2 and 7 are alternate interior angles, as well as angles 6 and 3. **Alternate exterior angles** are opposite angles formed by a transversal but, in contrast to interior angles, exterior angles exist outside the two original lines. Therefore, angles 1 and 8 are alternate exterior angles and so are angles 5 and 4. Finally, **consecutive interior angles** are pairs of angles formed by a transversal. These angles are located on the same side of the transversal and inside the two original lines. Therefore, angles 2 and 3 are a pair of consecutive interior angles, and so are angles 6 and 7. These definitions are instrumental in solving many problems that involve determining relationships between angles.

Determining the Measure of an Unknown Angle

To determine angle measures for adjacent angles, angles that share a common side and vertex, at least one of the angles must be known. Other information that is necessary to determine such measures include that there are 90^0 in a right angle, and there are 180^0 in a straight line. Therefore, if two adjacent angles form a right angle, they will add up to 90^0, and if two adjacent angles form a straight line, they add up to 180^0.

If the measurement of one of the adjacent angles is known, the other can be found by subtracting the known angle from the total number of degrees.

For example, given the following situation, if angle *a* measures 55^0, find the measure of unknown angle *b*:

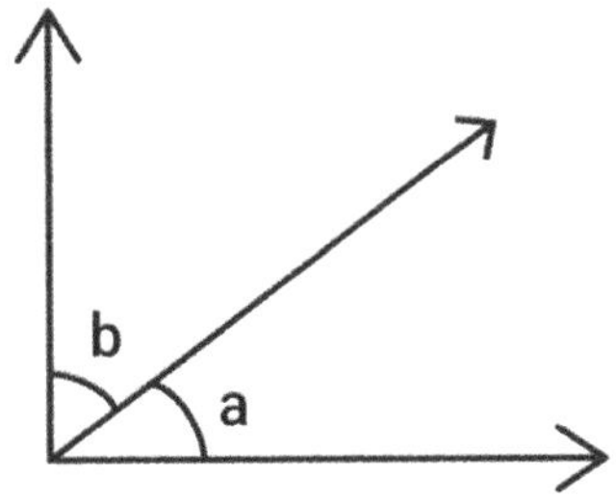

To solve this simply subtract the known angle measure from 90^0.

$$90° - 55° = 35°$$

The measure of $b = 35°$.

Given the following situation, if angle 1 measures 45^0, find the measure of the unknown angle 2:

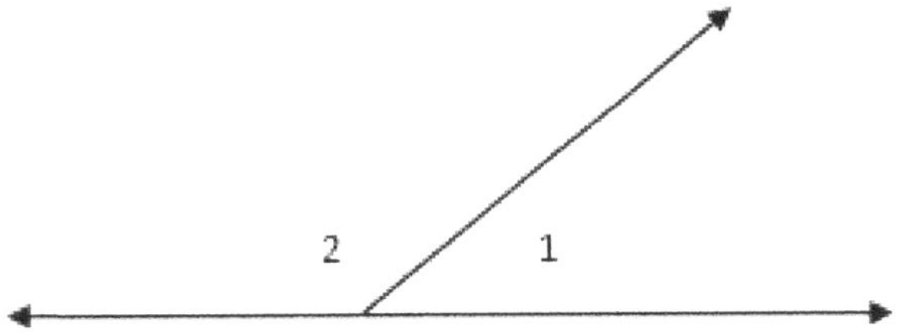

To solve this, simply subtract the known angle measure from 180^0.

$$180° - 45° = 135°$$

The measure of angle 2 = $135°$.

In the case that more than two angles are given, use the same method of subtracting the known angles from the total measure.

For example, given the following situation, if angle $y = 40^0$, and angle $z = 25^0$, find unknown angle x.

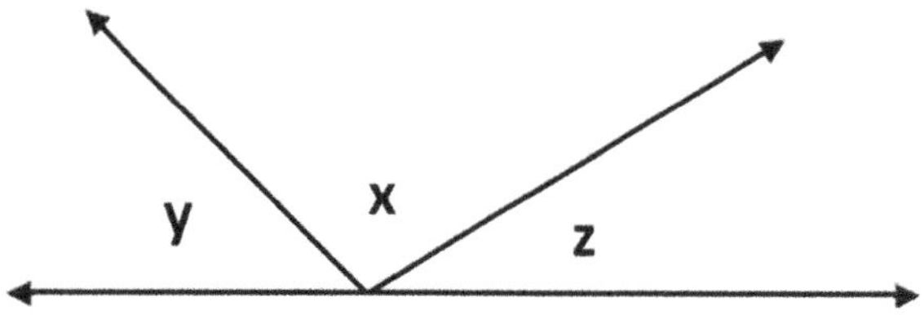

Subtract the known angles from 180^0.

$$180° - 40° - 25° = 115°$$

The measure of angle $x = 115^0$.

Solving Problems Involving Area, Volume, and Surface Area of Two- and Three-Dimensional Objects

Perimeter is the distance measurement around something. It can be thought of as the length of the boundary, like a fence. In contrast, area is the space occupied by a defined enclosure, like a field enclosed by a fence.

The perimeter of a square is measured by adding together all of the sides. Since a square has four equal sides, its perimeter can be calculated by multiplying the length of one side by 4.

Thus, the formula is $P = 4 \times s$, where s equals one side. The area of a square is calculated by squaring the length of one side, which is expressed as the formula $A = s^2$.

Like a square, a rectangle's perimeter is measured by adding together all of the sides. But as the sides are unequal, the formula is different. A rectangle has equal values for its lengths (long sides) and equal values for its widths (short sides), so the perimeter formula for a rectangle is $P = l + l + w + w = 2l + 2w$, where l equals length and w equals width. The area is found by multiplying the length by the width, so the formula is $A = l \times w$.

A triangle's perimeter is measured by adding together the three sides, so the formula is $P = a + b + c$, where $a, b,$ and c are the values of the three sides. The area is calculated by multiplying the length of the base times the height times ½, so the formula is:

$$A = \frac{1}{2} \times b \times h = \frac{bh}{2}$$

The base is the bottom of the triangle, and the height is the distance from the base to the peak. If a problem asks one to calculate the area of a triangle, it will provide the base and height.

A circle's perimeter—also known as its **circumference**—is measured by multiplying the **diameter** (the straight line measured from one side, through the center, to the direct opposite side of the circle) by π, so the formula is $\pi \times d$. This is sometimes expressed by the formula $C = 2 \times \pi \times r$, where r is the **radius** of the circle. These formulas are equivalent, as the radius equals half of the diameter. The area of a circle is calculated with the formula:

$$A = \pi \times r^2$$

The test will indicate either to leave the answer with π attached or to calculate to the nearest decimal place, which means multiplying by 3.14 for π.

The perimeter of a parallelogram is measured by adding the lengths and widths together. Thus, the formula is the same as for a rectangle:

$$P = l + l + w + w = 2l + 2w$$

However, the area formula differs from the rectangle. For a parallelogram, the area is calculated by multiplying the length by the height: $A = h \times l$

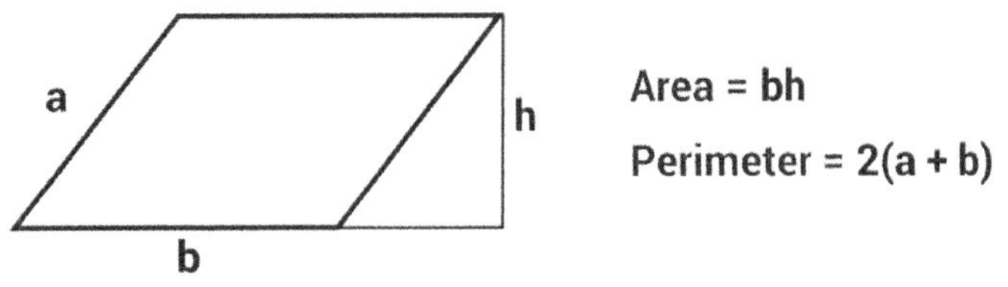

The perimeter of a trapezoid is calculated by adding the two unequal bases and two equal sides, so the formula is:

$$P = a + b_1 + c + b_2$$

Although unlikely to be a test question, the formula for the area of a trapezoid is $A = \frac{b_1 + b_2}{2} \times h$, where h equals height, and b_1 and b_2 equal the bases.

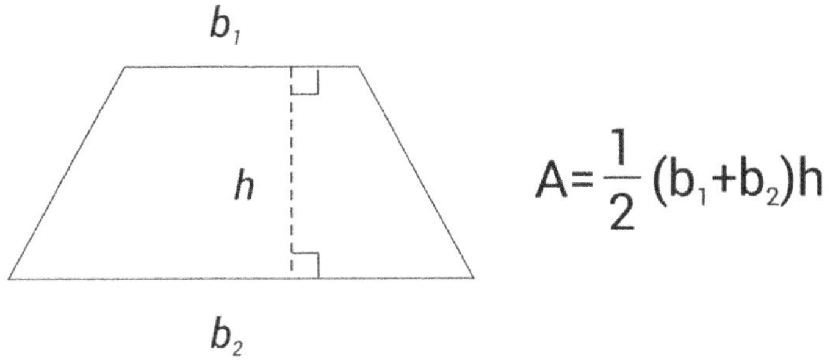

Using formulas such as perimeter and area for different shapes, it's possible to solve for missing side lengths.

Consider the following problem:

The total perimeter of a rectangular garden is 36 m. If the length of each side is 12 m, what is the width?

The formula for the perimeter of a rectangle is P=2L+2W, where P is the perimeter, L is the length, and W is the width. The first step is to substitute all of the data into the formula:

$$36 = 2(12) + 2W$$

Simplify by multiplying 2x12:

$$36 = 24 + 2W$$

Simplifying this further by subtracting 24 on each side, which gives:

$$36-24 = 24-24+2W$$

$$12= 2W$$

Divide by 2:

$$6 = W$$

The width is 6 m. Remember to test this answer by substituting this value into the original formula:

$$36 = 2(12) + 2(6)$$

More complicated situations can arise where missing side lengths can be calculated by using concepts of similarity and proportional relationships. Suppose that Lara is 5 feet tall and is standing 30 feet from the base of a light pole, and her shadow is 6 feet long. How high is the light on the pole? To figure this out, it helps to make a sketch of the situation:

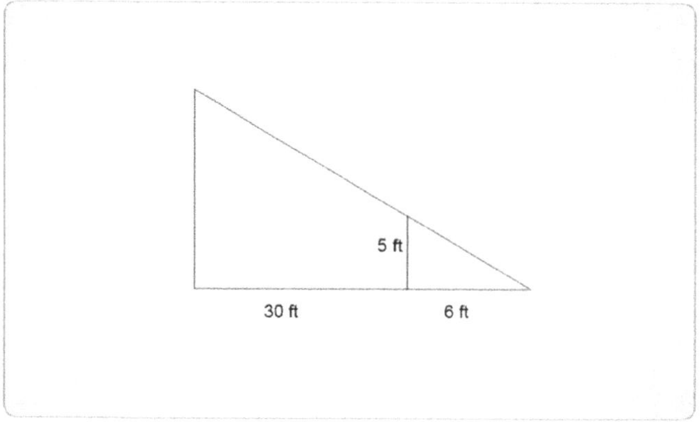

The light pole is the left side of the triangle. Lara is the 5-foot vertical line. Test takers should notice that there are two right triangles here, and that they have all the same angles as one another. Therefore, they form similar triangles. So, the ratio of proportionality between them must be found.

The bases of these triangles are known. The small triangle, formed by Lara and her shadow, has a base of 6 feet. The large triangle, formed by the light pole along with the line from the base of the pole out to the end of Lara's shadow is $30 + 6 = 36$ feet long. So, the ratio of the big triangle to the little triangle is $\frac{36}{6} = 6$. The height of the little triangle is 5 feet. Therefore, the height of the big triangle will be $6 \cdot 5 = 30$ feet, meaning that the light is 30 feet up the pole.

Composite Shapes

The perimeter of an irregular polygon is found by adding the lengths of all of the sides. In cases where all of the sides are given, this will be very straightforward, as it will simply involve finding the sum of the provided lengths. Other times, a side length may be missing and must be determined before the perimeter can be calculated.

Consider the example below:

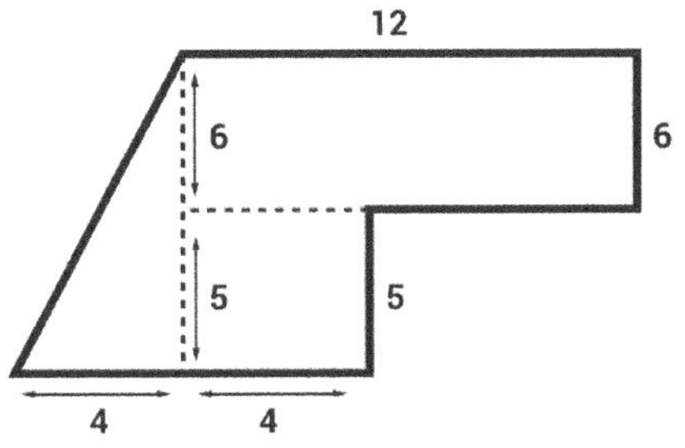

All of the side lengths are provided except for the angled side on the left. Test takers should notice that this is the hypotenuse of a right triangle. The other two sides of the triangle are provided (the base is 4 and the height is 6 + 5 = 11). The Pythagorean Theorem can be used to find the length of the hypotenuse, remembering that $a^2 + b^2 = c^2$.

Substituting the side values provided yields $(4)^2 + (11)^2 = c^2$.

Therefore, $c = \sqrt{16 + 121} = 11.7$

Finally, the perimeter can be found by adding this new side length with the other provided lengths to get the total length around the figure:

$$4 + 4 + 5 + 8 + 6 + 12 + 11.7 = 50.7$$

Although units are not provided in this figure, remember that reporting units with a measurement is important.

The area of irregular polygons is found by decomposing, or breaking apart, the figure into smaller shapes. When the area of the smaller shapes is determined, the area of the smaller shapes will produce the area of the original figure when added together.

Consider the earlier example:

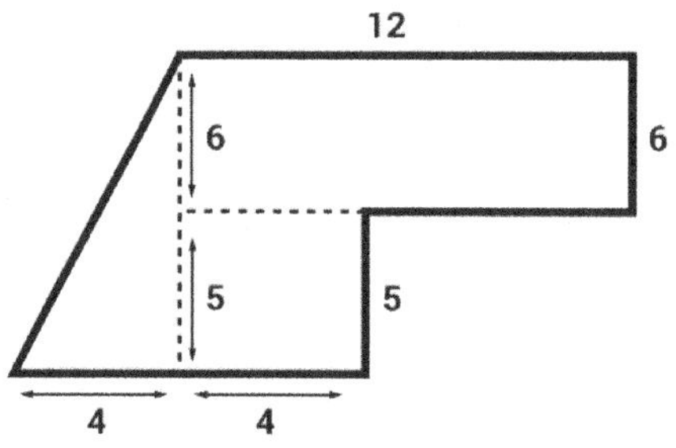

The irregular polygon is decomposed into two rectangles and a triangle. The area of the large rectangles ($A = l \times w \rightarrow A = 12 \times 6$) is 72 square units.

The area of the small rectangle is 20 square units ($A = 4 \times 5$). The area of the triangle ($A = \frac{1}{2} \times b \times h \rightarrow A = \frac{1}{2} \times 4 \times 11$) is 22 square units. The sum of the areas of these figures produces the total area of the original polygon:

$$A = 72 + 20 + 22 \rightarrow A = 114 \text{ square units}$$

Here's another example:

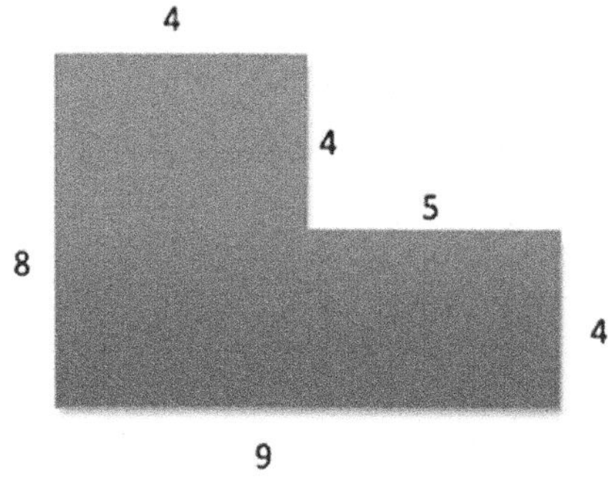

This irregular polygon is decomposed into two rectangles. The area of the large rectangle:

$$(A = l \times w \rightarrow A = 8 \times 4) \text{ is 32 square units}$$

The area of the small rectangle is 20 square units ($A = 4 \times 5$). The sum of the areas of these figures produces the total area of the original polygon:

$$A = 32 + 20 \rightarrow A = 52 \text{ square units}$$

Measuring Volumes

Volume is the measurement of how much space an object occupies, like how much space is in the cube. Volume is useful in determining the space within a certain three-dimensional object. Volume can be calculated for a cube, rectangular prism, cylinder, pyramid, cone, and sphere. By knowing specific dimensions of the objects, the volume of the object is computed with these figures. The units for the volumes of solids can include cubic centimeters, cubic meters, cubic inches, and cubic feet.

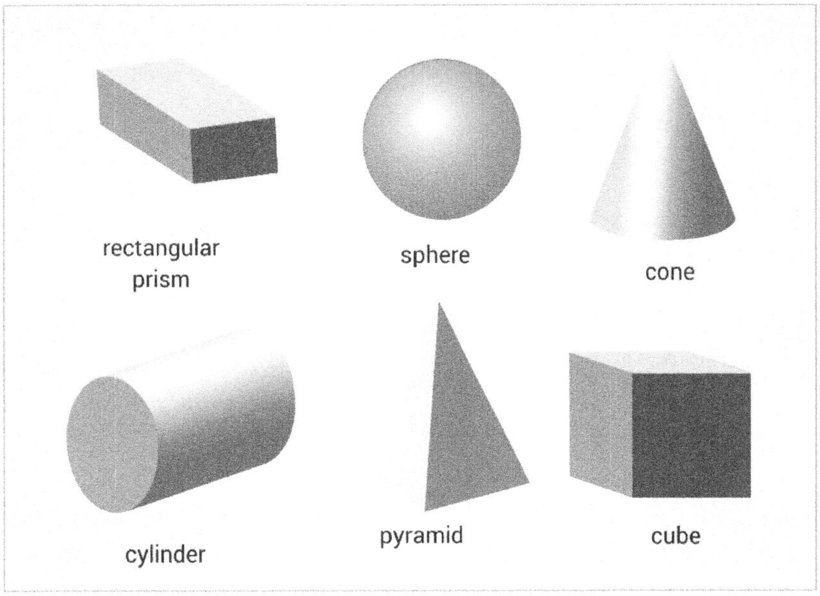

Cube

The cube is the simplest figure for which volume can be determined because all dimensions in a cube are equal. In the following figure, the length, width, and height of the cube are all represented by the variable *a* because these measurements are equal lengths.

The volume of any rectangular, three-dimensional object is found by multiplying its length by its width by its height. In the case of a cube, the length, width, and height are all equal lengths, represented by the variable *a*. Therefore, the equation used to calculate the volume is $(a \times a \times a)$ or a^3. In a real-world example of this situation, if the length of a side of the cube is 3 centimeters, the volume is calculated by utilizing the formula $(3 \times 3 \times 3) = 9$ cm^3.

Rectangular Prism

A rectangular prism is a box whose sides are all rectangles meeting at 90° angles. Such a box has three dimensions: length, width, and height. The dimensions of a rectangular prism are not necessarily equal as those of a cube. Therefore, the formula for a rectangular prism recognizes that the dimensions vary and use different variables to represent these lengths. The length, width, and height of a rectangular prism are represented with the variables *l, w*, and *h*.

The equation used to calculate volume is length times width times height. Using the variables in the diagram above, this means $l \times w \times h$. In a real-world application of this situation, if *l*=2 cm*, w*=3 cm, and *h*=4 cm, the volume is calculated by utilizing the formula $3 \times 4 \times 5 = 60$ cm^3.

If we were to find the volume of two rectangular prisms, all we would need to do would be to find the total number of unit cubes within both regions. This would be the same as finding the individual volume of each prism and then adding them together because volume is considered **additive.** For instance, if you wanted to find the total volume of two of the prisms shown above, you would add 36 plus 36 to obtain 72 cubic centimeters. A situation in which something like this would be necessary would be when filling up fish tanks with salt water. Let's say you had two tanks with different dimensions, and you needed to find out how much salt water total was necessary to fill both tanks. You would find the individual volume of each tank, by counting unit cubes, and then adding them together.

The prisms do not have to be of equal size. For instance, consider any of the four following rectangular prisms:

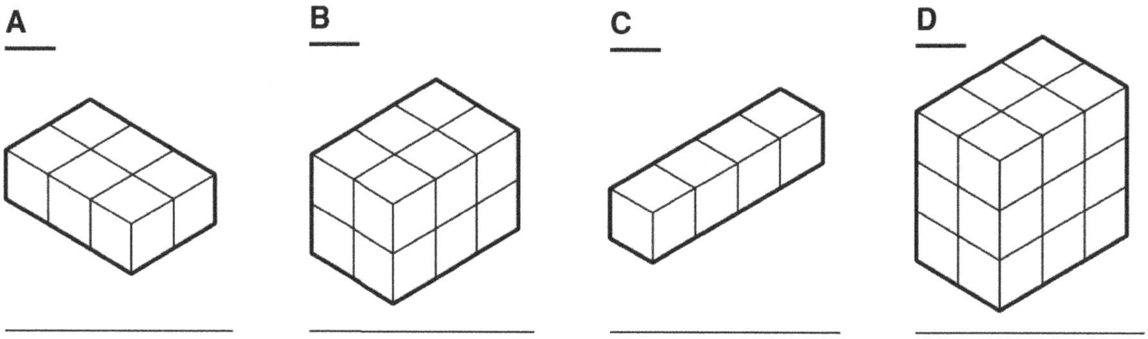

Prism a) has a volume of 6 cubic units, b) has a volume of 12 cubic units, c) has a volume of 5 cubic units and d) has a volume of 18 cubic units. Any of these could be formed with one another. For instance, prism a) could be laid on top of prism b) and the end result would have a volume of 18 cubic units because 6 + 12 = 18. Remember that volume has cubic units like cubic inches (in^3).

Pyramid

A **rectangular pyramid** is a figure with a rectangular base and four triangular sides that meet at a single vertex. If the rectangle has sides of lengths x and y, then the volume will be given by:

$$V = \frac{1}{3}xyh$$

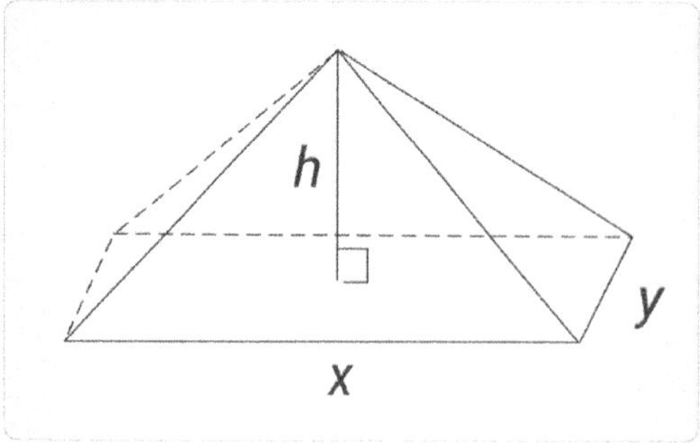

Cone

The formula to calculate the volume of a circular cone is similar to the formula for the volume of a pyramid. The primary difference in determining the area of a cone is that a circle serves as the base of a cone. Therefore, the area of a circle is used for the cone's base.

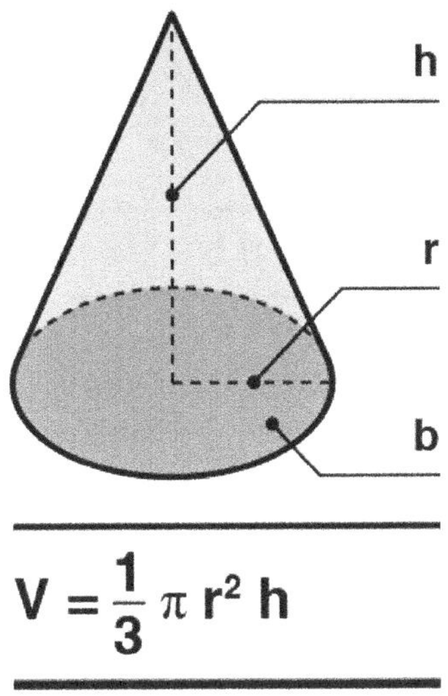

The variable *r* represents the radius, and the variable *h* represents the height of the cone. The formula used to calculate the volume of a cone is:

$$\frac{1}{3}\pi r^2 h$$

Essentially, the area of the base of the cone is multiplied by the cone's height. In a real-life example where the radius of a cone is 2 meters and the height of a cone is 5 meters, the volume of the cone is calculated by utilizing the formula:

$$\frac{1}{3}\pi 2^2 \times 5 = 21$$

After substituting 3.14 for π, the volume is 785.4 ft^3.

Sphere

The volume of a sphere uses π due to its circular shape. The length of the radius, *r*, is the only variable needed to determine the sphere's volume. The formula to calculate the volume of a sphere is $\frac{4}{3}\pi r^3$. Therefore, if the radius of a sphere is 8 centimeters, the volume of the sphere is calculated by utilizing the formula:

$$\frac{4}{3}\pi(8)^3 = 2{,}143\ cm^3$$

Cylinder

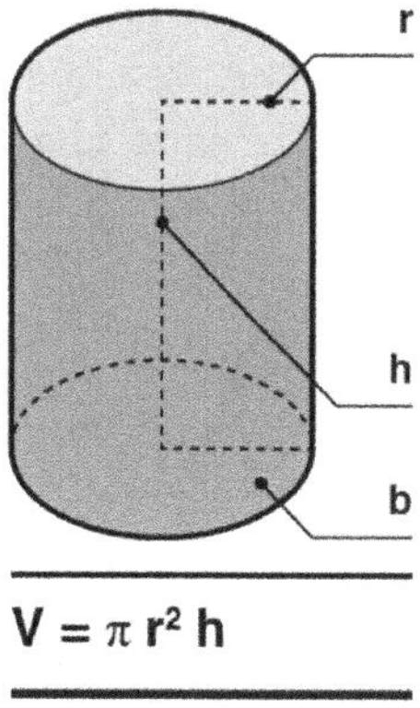

The formula to find the volume of a cylinder is $\pi r^2 h$. This formula contains the formula for the area of a circle (πr^2) because the base of a cylinder is a circle. To calculate the volume of a cylinder, the slices of circles needed to build the entire height of the cylinder are added together. For example, if the radius is 5 feet and the height of the cylinder is 10 feet, the cylinder's volume is calculated by using the following equation: $\pi 5^2 \times 10$. Substituting 3.14 for π, the volume is 785 ft^3.

Problems involving volume, length, and other units can also be solved using ratios. If the following graphic of a cone is given, the problem may ask for the volume to be found.

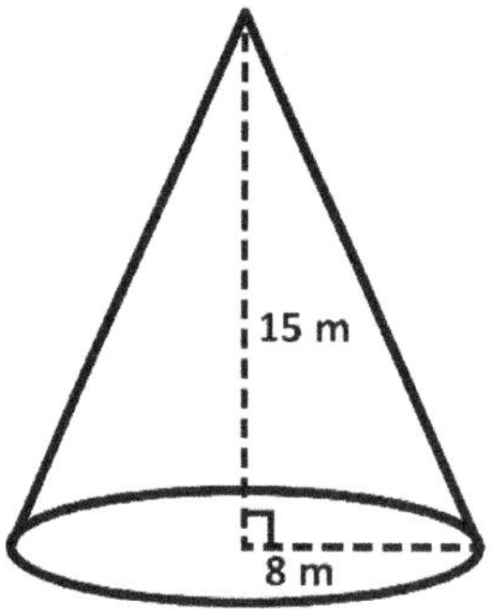

Referring to the formulas provided on the test, the volume of a cone is given as: $V = \pi r^2 \frac{h}{3}$, where r is the radius, and h is the height. Plugging $r = 8$ and $h = 15$ from the graphic into the formula, the following is obtained:

$$V = \pi(8^2)\frac{15}{3}$$

Therefore, volume of the cone is found to be $1005.3m^3$. Sometimes, answers in different units are sought. If this problem wanted the answer in liters, $1005.3m^3$ would need to be converted. Using the equivalence statement $1m^3$ = 1000L, the following ratio would be used to solve for liters: $1005.3\text{m}^3 \times \frac{1000L}{1m^3}$. Cubic meters in the numerator and denominator cancel each other out, and the answer is converted to 1,005,300 liters, or 1.0053×10^6 L.

Other conversions can also be made between different given and final units. If the temperature in a pool is 30°C, what is the temperature of the pool in degrees Fahrenheit? To convert these units, an equation is used relating Celsius to Fahrenheit. The following equation is used: $T_{°F} = 1.8T_{°C} + 32$. Plugging in the given temperature and solving the equation for T yields the result: $T_{°F} = 1.8(30) + 32 = 86°F$. Both units in the metric system and U.S. customary system are widely used.

Surface Area

The area of a two-dimensional figure refers to the number of square units needed to cover the interior region of the figure. This concept is similar to wallpaper covering the flat surface of a wall. For example, if a rectangle has an area of 8 square inches (written 8 in^2), it will take 8 squares, each with sides one inch in length, to cover the interior region of the rectangle. Note that area is measured in square units such as: square feet or ft^2; square yards or yd^2; square miles or mi^2.

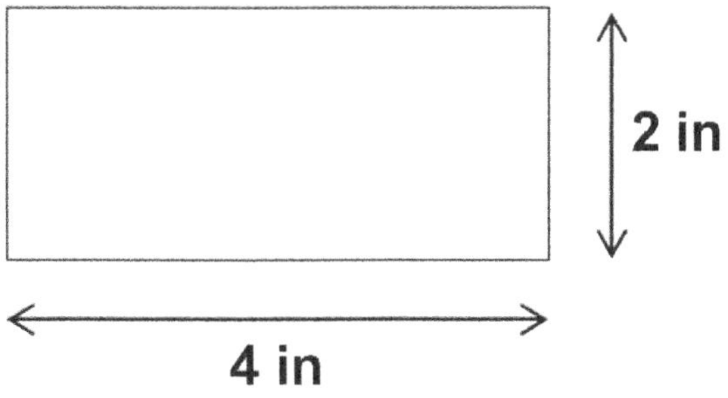

The surface area of a three-dimensional figure refers to the number of square units needed to cover the entire surface of the figure. This concept is similar to using wrapping paper to completely cover the outside of a box. For example, if a triangular pyramid has a surface area of 17 square inches (written $17in^2$), it will take 17 squares, each with sides one inch in length, to cover the entire surface of the pyramid. Surface area is also measured in square units.

The formula for a cube's surface area is $SA = 6 \times s^2$, where s is the length of a side. A cube has 6 equal sides, so the formula expresses the area of all the sides. Volume is simply measured by taking the cube of the length, so the formula is $V = s^3$.

The surface area formula for a rectangular prism or a general box is SA = $2(lw + lh + wh)$, where l is the length, h is the height, and w is the width. The volume formula is $V = l \times w \times h$, which is the cube's volume formula adjusted for the unequal lengths of a box's sides.

The formula for a sphere's surface area is $SA = 4\pi r^2$, where r is the sphere's radius. The surface area formula is the area for a circle multiplied by four.

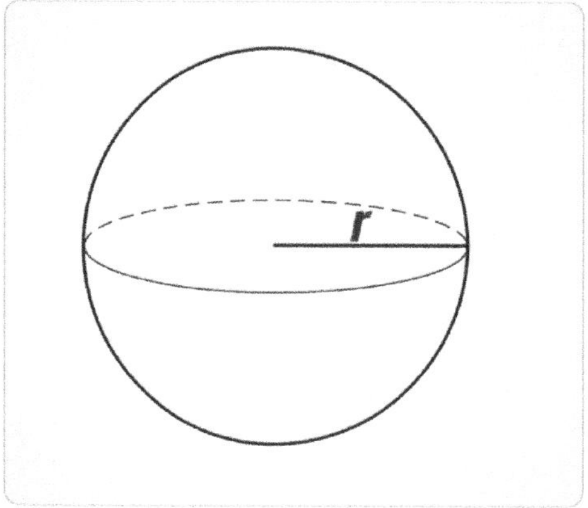

Many three-dimensional figures (solid figures) can be represented by nets consisting of rectangles and triangles. The surface area of such solids can be determined by adding the areas of each of its faces and bases. Finding the surface area using this method requires calculating the areas of rectangles and triangles. To find the area (*A*) of a rectangle, the length (*l*) is multiplied by the width:

$$(w) \rightarrow A = l \times w$$

The area of a rectangle with a length of 8cm and a width of 4cm is calculated:

$$A = (8cm) \times (4cm) \rightarrow A = 32cm^2$$

To calculate the area (*A*) of a triangle, the product of $\frac{1}{2}$, the base (*b*), and the height (*h*) is found

$$A = \frac{1}{2} \times b \times h$$

Note that the height of a triangle is measured from the base to the vertex opposite of it forming a right angle with the base. The area of a triangle with a base of 11cm and a height of 6cm is calculated:

$$A = \frac{1}{2} \times (11cm) \times (6cm) \rightarrow A = 33cm^2$$

Consider the following triangular prism, which is represented by a net consisting of two triangles and three rectangles.

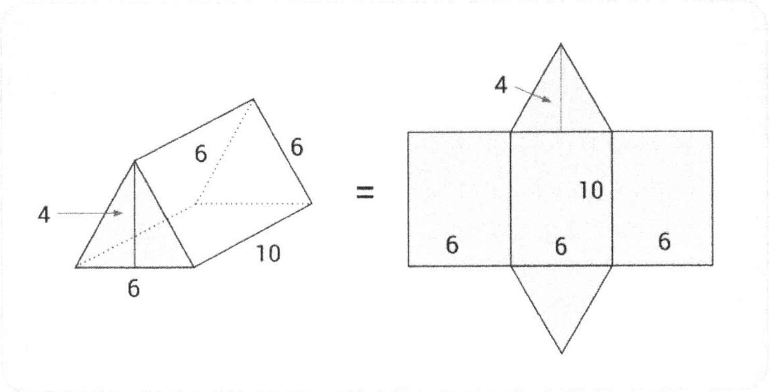

The surface area of the prism can be determined by adding the areas of each of its faces and bases. The surface area (*SA*) = area of triangle + area of triangle + area of rectangle + area of rectangle + area of rectangle.

$$SA = \left(\frac{1}{2} \times b \times h\right) + \left(\frac{1}{2} \times b \times h\right) + (l \times w) + (l \times w) + (l \times w)$$

$$SA = \left(\frac{1}{2} \times 6 \times 4\right) + \left(\frac{1}{2} \times 6 \times 4\right) + (6 \times 10) + (6 \times 10) + (6 \times 10)$$

$$SA = (12) + (12) + (60) + (60) + (60)$$

$$SA = 204 \textit{ square units}$$

Statistics & Probability

Using Random Sampling to Draw Inferences About a Population

Using Data from a Random Sample to Draw Inferences About a Population With an Unknown Characteristic of Interest

The field of statistics describes relationships between quantities that are related, but not necessarily in a deterministic manner. For example, a graduating student's salary will often be higher when the student graduates with a higher GPA, but this is not always the case. Likewise, people who smoke tobacco are more likely to develop lung cancer, but, in fact, it is possible for non-smokers to develop the disease as well. **Statistics** describes these kinds of situations, where the likelihood of some outcome depends on the starting data.

Descriptive statistics involves analyzing a collection of data to describe its broad properties such average (or mean), what percent of the data falls within a given range, and other such properties. An example of this would be taking all of the test scores from a given class and calculating the average test score. **Inferential statistics** attempts to use data about a subset of some population to make inferences about the rest of the population. An example of this would be taking a collection of students who received tutoring and comparing their results to a collection of students who did not receive tutoring, then using that comparison to try to predict whether the tutoring program in question is beneficial.

To be sure that inferences have a high probability of being true for the whole population, the subset that is analyzed needs to resemble a miniature version of the population as closely as possible. For this reason, statisticians like to choose random samples from the population to study, rather than picking a specific group of people based on some similarity. For example, studying the incomes of people who live in Portland does not tell anything useful about the incomes of people who live in Tallahassee.

A statistical question is answered by collecting data with variability. Data consists of facts and/or statistics (numbers), and variability refers to a tendency to shift or change. Data is a broad term, inclusive of things like height, favorite color, name, salary, temperature, gas mileage, and language. Questions requiring data as an answer are not necessarily statistical questions. If there is no variability in the data, then the question is not statistical in nature. Consider the following examples: what is Mary's favorite color? How much money does your mother make? What was the highest temperature last week? How many miles did your car get on its last tank of gas? How much taller than Bob is Ed?

None of the above are statistical questions because each case lacks variability in the data needed to answer the question. The questions on favorite color, salary, and gas mileage each require a single piece of data, whether a fact or statistic. Therefore, variability is absent. Although the temperature question requires multiple pieces of data (the high temperature for each day), a single, distinct number is the answer. The height question requires two pieces of data, Bob's height and Ed's height, but no difference in variability exists between those two values. Therefore, this is not a statistical question. Statistical questions typically require calculations with data.

Consider the following statistical questions:

How many miles per gallon of gas does the 2016 Honda Civic get? To answer this question, data must be collected. This data should include miles driven and gallons used. Different cars, different drivers, and different driving conditions will produce different results. Therefore, variability exists in the data. To answer the question, the mean (average) value could be determined.

Are American men taller than Chinese men? To answer this question, data must be collected. This data should include the heights of American men and the heights of Chinese men. All American men are not the same height and all Chinese men are not the same height. Some American men are taller than some Chinese men and some Chinese men are taller than some American men. Therefore, variability exists in the data. To answer the question, the median values for each group could be determined and compared.

The following are more examples of statistical questions: What proportion of 4th graders have a favorite color of blue? How much money do teachers make? Is it colder in Boston or Chicago?

Statistics involves making decisions and predictions about larger data sets based on smaller data sets. Basically, the information from one part or subset can help predict what happens in the entire data set or population at large. The entire process involves guessing, and the predictions and decisions may not be 100 percent correct all of the time; however, there is some truth to these predictions, and the decisions do have mathematical support. The smaller data set is called a **sample** and the larger data set (in which the decision is being made) is called a **population.** A **random sample** is used as the sample, which is an unbiased collection of data points that represents the population as well as it can. There are many methods of forming a random sample, and all adhere to the fact that every potential data point has a predetermined probability of being chosen. Statistical inference, based in probability theory, makes calculated assumptions about an entire population based on data from a sample set from that population.

A population is the entire set of people or things of interest. Suppose a study is intended to determine the number of hours of sleep per night for college females in the U.S. The population would consist of EVERY college female in the country. A sample is a subset of the population that may be used for the study. It would not be practical to survey every female college student, so a sample might consist of 100 students per school from 20 different colleges in the country. From the results of the survey, a sample statistic can be calculated. A sample statistic is a numerical characteristic of the sample data, including mean and variance. A sample statistic can be used to estimate a corresponding population parameter. A population parameter is a numerical characteristic of the entire population. Suppose the sample data had a mean (average) of 5.5. This sample statistic can be used as an estimate of the population parameter (average hours of sleep for every college female in the U.S.).

The presentation of statistics can be manipulated to produce a desired outcome. For example, in the statement "four out of five dentists recommend our toothpaste", critical readers should wonder: *who are the five dentists?* While the wording is similar, this statement is very different from "four out of every five dentists recommend our toothpaste." The context of the numerical values allows one to decipher the meaning, intent, and significance of the survey or study.

When analyzing a report, the researchers who conducted the study and their intent must be considered. Was it performed by a neutral party or by a person or group with a vested interest? The sampling method and the data collection method should also be evaluated. Was it a true random sample of the population or was one subgroup over- or underrepresented? Lastly, the measurement system used to obtain the data should be assessed. Was the system accurate and precise or was it a flawed system?

Using Statistics to Gain Information About a Population by Examining a Sample of the Population

A population parameter is usually unknown and therefore is estimated using a sample statistic. This estimate may be highly accurate or relatively inaccurate based on errors in sampling. A confidence interval indicates a range of values likely to include the true population parameter. These are constructed at a given confidence level, such as 95%. This means that if the same population is sampled repeatedly, the true population parameter would occur within the interval for 95% of the samples.

The accuracy of a population parameter based on a sample statistic may also be affected by measurement error, which is the difference between a quantity's true value and its measured value. Measurement error can be divided into random error and systematic error. An example of random error for the previous scenario would be a student reporting 8 hours of sleep when she actually sleeps 7 hours per night. Systematic errors are those attributed to the measurement system. Suppose the sleep survey gave response options of 2, 4, 6, 8, or 10 hours. This would lead to systematic measurement error.

Drawing Informal Comparative Inferences about Two Populations

Using Measures of Center and Variability for Numerical Data From Random Samples to Draw Informal Comparative Inferences About Two Populations

Comparing data sets within statistics can mean many things. The first way to compare data sets is by looking at the center and spread of each set. The center of a data set is measured by mean, median, and mode. These are sometimes referred to as measures of central tendency.

Mean

The first property that can be defined for this set of data is the **mean**. This is the same as average. To find the mean, add up all the data points, then divide by the total number of data points. For example, suppose that in a class of 10 students, the scores on a test were 50, 60, 65, 65, 75, 80, 85, 85, 90, 100. Therefore, the average test score will be:

$$\frac{50 + 60 + 65 + 65 + 75 + 80 + 85 + 85 + 90 + 100}{10} = 75.5$$

The mean is a useful number if the distribution of data is normal (more on this later), which roughly means that the frequency of different outcomes has a single peak and is roughly equally distributed on both sides of that peak. However, it is less useful in some cases where the data might be split or where there are some outliers. **Outliers** are data points that are far from the rest of the data. For example, suppose there are 10 executives and 90 employees at a company. The executives make \$1000 per hour, and the employees make \$10 per hour.

Therefore, the average pay rate will be:

$$\frac{\$1000 \times 10 + \$10 \times 90}{100} = \$109\ per\ hour$$

In this case, this average is not very descriptive since it's not close to the actual pay of the executives *or* the employees.

Median

Another useful measurement is the **median**. In a data set, the median is the point in the middle. The middle refers to the point where half the data comes before it and half comes after, when the data is recorded in numerical order. For instance, these are the speeds of the fastball of a pitcher during the last inning that he pitched (in order from least to greatest):

90, 92, 93, 93, 95, 96, 97, 97, 97

There are nine total numbers, so the middle or *median* number in the 5th one, which is 95.

In cases where the number of data points is an even number, then the average of the two middle points is taken. In the previous example of test scores, the two middle points are 75 and 80. Since there is no single point, the average of these two scores needs to be found. The average is:

$$\frac{75+80}{2} = 77.5$$

The median is generally a good value to use if there are a few outliers in the data. It prevents those outliers from affecting the "middle" value as much as when using the mean.

Since an outlier is a data point that is far from most of the other data points in a data set, this means an outlier also is any point that is far from the median of the data set. The outliers can have a substantial effect on the mean of a data set, but usually do not change the median or mode, or do not change them by a large quantity. For example, consider the data set (3, 5, 6, 6, 6, 8). This has a median of 6 and a mode of 6, with a mean of $\frac{34}{6} \approx 5.67$. Now, suppose a new data point of 1000 is added so that the data set is now (3, 5, 6, 6, 6, 8, 1000). This does not change the median or mode, which are both still 6. However, the average is now $\frac{1034}{7}$, which is approximately 147.7. In this case, the median and mode will be better descriptions for most of the data points.

The reason for outliers in a given data set is a complicated problem. It is sometimes the result of an error by the experimenter, but often they are perfectly valid data points that must be taken into consideration.

Mode

One additional measure to define for X is the **mode**. This is the data point that appears most frequently. If two or more data points all tie for the most frequent appearance, then each of them is considered a mode. In the case of the test scores, where the numbers were 50, 60, 65, 65, 75, 80, 85, 85, 90, 100, there are two modes: 65 and 85.

Spread

Methods for determining the **spread** of the sample include calculating the range and standard deviation for the data. The **range** is calculated by subtracting the lowest value from the highest value in the set. Given a data set X consisting of data points $(x_1, x_2, x_3, \dots x_n)$, the **variance** of X is defined to be:

$$\frac{\sum_{i=1}^{n}(x_i - \bar{X})^2}{n}$$

This means that the variance of X is the average of the squares of the differences between each data point and the mean of X.

Given a data set X consisting of data points $(x_1, x_2, x_3, \ldots x_n)$, the **standard deviation** of X is defined to be:

$$s_X = \sqrt{\frac{\sum_{i=1}^{n}(x_i - \bar{X})^2}{n}}$$

In other words, the standard deviation is the square root of the variance.

Both the variance and the standard deviation are measures of how much the data tend to be spread out. When the standard deviation is low, the data points are mostly clustered around the mean. When the standard deviation is high, this generally indicates that the data are quite spread out, or else that there are a few substantial outliers.

As a simple example, compute the standard deviation for the data set (1, 3, 3, 5). First, compute the mean, which will be:

$$\frac{1+3+3+5}{4} = \frac{12}{4} = 3$$

Now, find the variance of X with the formula:

$$\sum_{i=1}^{4}(x_i - \bar{X})^2 = (1-3)^2 + (3-3)^2 + (3-3)^2 + (5-3)^2$$

$$-2^2 + 0^2 + 0^2 + 2^2 = 8$$

Therefore, the variance is $\frac{8}{4} = 2$. Taking the square root, the standard deviation will be $\sqrt{2}$.

Note that the standard deviation only depends upon the mean, not upon the median or mode(s). Generally, if there are multiple modes that are far apart from one another, the standard deviation will be high. A high standard deviation does not always mean there are multiple modes, however.

Quartiles and Percentiles

The **first quartile** of a set of data X refers to the largest value from the first ¼ of the data points. In practice, there are sometimes slightly different definitions that can be used, such as the median of the first half of the data points (excluding the median itself if there are an odd number of data points). The term also has a slightly different use: when it is said that a data point lies *in the first quartile*, it means it is less than or equal to the median of the first half of the data points. Conversely, if it lies *at* the first quartile, then it is equal to the first quartile.

When it is said that a data point lies in the **second quartile**, it means it is between the first quartile and the median.

The **third quartile** refers to data that lies between ½ and ¾ of the way through the data set. Again, there are various methods for defining this precisely, but the simplest way is to include all of the data that lie between the median and the median of the top half of the data.

Data that lies in the **fourth quartile** refers to all of the data above the third quartile.

Percentiles may be defined in a similar manner to quartiles. Generally, this is defined in the following manner:

If a data point lies **in the n-th percentile**, this means it lies in the range of the first *n*% of the data.

If a data point lies **at the n-th percentile**, then it means that *n*% of the data lies below this data point.

<u>Describing a Set of Data</u>
A set of data can be described in terms of its center, spread, shape and any unusual features. The center of a data set can be measured by its mean, median, or mode. The spread of a data set refers to how far the data points are from the center (mean or median). A data set with data points clustered around the center will have a small spread. A data set covering a wide range will have a large spread.

When a data set is displayed as a graph like the one below, the shape indicates if a sample is normally distributed, symmetrical, or has measures of skewness. When graphed, a data set with a normal distribution will resemble a bell curve.

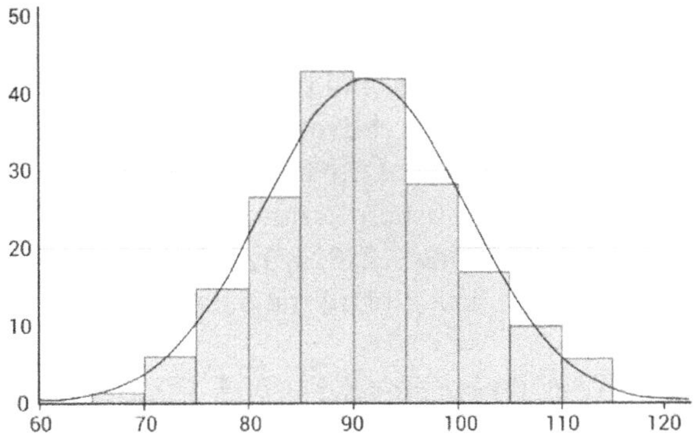

If the data set is symmetrical, each half of the graph when divided at the center is a mirror image of the other. If the graph has fewer data points to the right, the data is skewed right. If it has fewer data points to the left, the data is skewed left.

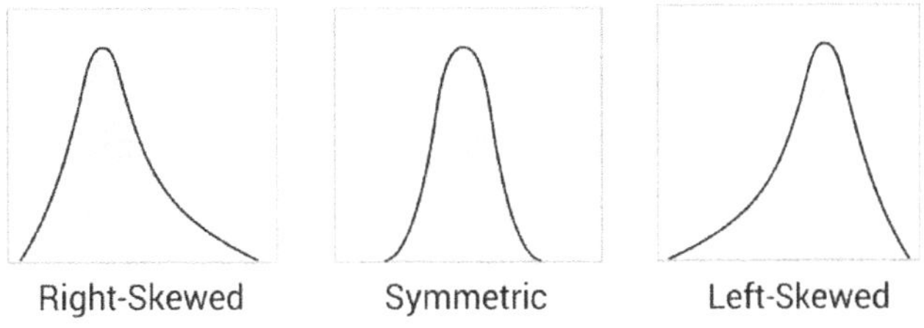

A description of a data set should include any unusual features such as gaps or outliers. A gap is a span within the range of the data set containing no data points. An outlier is a data point with a value either extremely large or extremely small when compared to the other values in the set.

The graphs above can be referred to as **unimodal** since they all have a single peak. This is contrast to **bimodal** graph that have multiple peaks.

Assessing the Degree of Visual Overlap of Two Numerical Data Distributions

Comparing data sets within statistics can mean many things. As mentioned, one of these ways is to look at the center and spread of each set. The center of a data set can mean two things: median or mean. The **median** is the value that's halfway into each data set, and it splits the data into two intervals. The **mean** is the average value of the data within a set. It's calculated by adding up all of the data in the set and dividing the total by the number of data points. Outliers can significantly impact the mean. Additionally, two completely different data sets can have the same mean. For example, a data set with values ranging from 0 to 100 and a data set with values ranging from 44 to 56 can both have means of 50. The first data set has a much wider range, which is known as the **spread** of the data. This measures how varied the data is within each set.

Another measure of variability that can be used when comparing two different data sets is the mean absolute deviation of a data set. This value is defined to be the average value of the distances from each data value to the mean of the set. Therefore, in order to find this quantity, first you must calculate the mean of the data set by adding up all of the values and dividing by the total number of values. Then, calculate the absolute value of the difference between each value and the found mean. Then, find the average of those values.

Imagine there are 8 students in your history class, and their scores on last week's test were 99, 95, 91, 88, 86, 84, 77, and 76 out of 100. In order to find the mean absolute deviation of this data set, first we need to find the average of the values. The sum of the 8 test scores is 696. Divide this by 8 to get an average test score of 87. Next, calculate the absolute value of the difference between each test score and the average. Those values are the following: 12, 8, 4, 1, 1, 3, 10, and 11. The average of those 8 values is 6.25, which is the mean absolute deviation of the data set.

Investigating Chance Processes and Developing, Using, and Evaluating Probability Models

The Probability of a Chance Event is a Number between 0 and 1

Probability is a measure of how likely an event is to occur. Probability is written as a fraction between zero and one. If an event has a probability of zero, the event will never occur. If an event has a probability of one, the event will definitely occur. If the probability of an event is closer to zero, the event is unlikely to occur. If the probability of an event is closer to one, the event is more likely to occur. For example, a probability of $\frac{1}{2}$ means that the event is equally as likely to occur as it is not to occur. An example of this is tossing a coin. To calculate the probability of an event, the number of favorable outcomes is divided by the number of total outcomes. For example, suppose you have 2 raffle tickets out of 20 total tickets sold. The probability that you win the raffle is calculated:

$$\frac{number\ of\ favorable\ outcomes}{total\ numberof\ outcomes} = \frac{2}{20} = \frac{1}{10} \text{(always reduce fractions)}$$

Therefore, the probability of winning the raffle is $\frac{1}{10}$ or 0.1.

Approximating the Probability of a Chance Event

Chance is the measure of how likely an event is to occur, written as a percent. If an event will never occur, the event has a 0% chance. If an event will certainly occur, the event has a 100% chance. If an event will sometimes occur, the event has a chance somewhere between 0% and 100%. To calculate chance, probability is calculated and the fraction is converted to a percent.

The probability of multiple events occurring can be determined by multiplying the probability of each event. For example, suppose you flip a coin with heads and tails, and roll a six-sided dice numbered one through six. To find the probability that you will flip heads AND roll a two, the probability of each event is determined and those fractions are multiplied. The probability of flipping heads is $\frac{1}{2}\left(\frac{1\ side\ with\ heads}{2\ sides\ total}\right)$ and the probability of rolling a two is $\frac{1}{6}\left(\frac{1\ side\ with\ a\ 2}{6\ total\ sides}\right)$. The probability of flipping heads AND rolling a 2 is:

$$\frac{1}{2} \times \frac{1}{6} = \frac{1}{12}$$

The above scenario with flipping a coin and rolling a dice is an example of independent events. Independent events are circumstances in which the outcome of one event does not affect the outcome of the other event. Conversely, dependent events are ones in which the outcome of one event affects the outcome of the second event. Consider the following scenario: a bag contains 5 black marbles and 5 white marbles. What is the probability of picking 2 black marbles without replacing the marble after the first pick?

The probability of picking a black marble on the first pick is:

$$\frac{5}{10}\left(\frac{5\ black\ marbles}{10\ total\ marbles}\right)$$

Assuming that a black marble was picked, there are now 4 black marbles and 5 white marbles for the second pick. Therefore, the probability of picking a black marble on the second pick is:

$$\frac{4}{9}\left(\frac{4\ black\ marbles}{9\ total\ marbles}\right)$$

To find the probability of picking two black marbles, the probability of each is multiplied:

$$\frac{5}{10} \times \frac{4}{9} = \frac{20}{90} = \frac{2}{9}$$

Developing a Probability Model to Find Probabilities of Events

Developing a Uniform Probability Model

A **uniform probability model** is one where each outcome has an equal chance of occurring, such as the probabilities of rolling each side of a die. A **non-uniform probability model** is one where each outcome has an unequal chance of occurring. In a uniform probability model, the conditional probability formulas for $P(B|A)$ and $P(A|B)$ can be multiplied by their respective denominators to obtain two formulas for $P(A \text{ and } B)$. Therefore, the multiplication rule is derived as:

$$P(A \text{ and } B) = P(A)P(B|A) = P(B)P(A|B)$$

In a model, if the probability of either individual event is known and the corresponding conditional probability is known, the multiplication rule allows the probability of the joint occurrence of A and B to be calculated.

Developing a Probability Model By Observing Frequencies in Data Generated From a Chance Process

Data that isn't described using numbers is known as **categorical data.** For example, age is numerical data, but hair color is categorical data. Categorical data is summarized using two-way frequency tables. A **two-way frequency table** counts the relationship between two sets of categorical data. There are rows and columns for each category, and each cell represents frequency information that shows the actual data count between each combination.

For example, the graphic on the left-side below is a two-way frequency table showing the number of girls and boys taking language classes in school. Entries in the middle of the table are known as the **joint frequencies**. For example, the number of girls taking French class is 12, which is a joint frequency. The totals are the **marginal frequencies**. For example, the total number of boys is 20, which is a marginal frequency. If the frequencies are changed into percentages based on totals, the table is known as a **two-way relative frequency table.** Percentages can be calculated using the table total, the row totals, or the column totals.

Here's the process of obtaining the two-way relative frequency table using the table total:

Two-Way Frequency Table

	Spanish	French	German	Total
Boys	10	2	8	**20**
Girls	15	12	3	**30**
Total	**25**	**14**	**11**	**50**

Joint frequencies

Marginal frequencies

Divide all tables entries by the table total (50) →

Two-Way Relative Frequency Table (with respect to table total)

	Spanish	French	German	Total
Boys	0.2	0.04	0.16	**0.40**
Girls	0.3	0.24	0.06	**0.60**
Total	**0.5**	**0.28**	**0.22**	**1.00**

Two-Way Frequency Table

	Spanish	French	German	Total
Boys	10	2	8	**20**
Girls	15	12	3	**30**
Total	**25**	**14**	**11**	**50**

Divide each column entry by that column's total →

Two-Way Relative Frequency Table (with respect to table total)

	Spanish	French	German	Total
Boys	0.4	0.14	0.73	**0.4**
Girls	0.6	0.86	0.27	**0.6**
Total	**1.00**	**1.00**	**1.00**	**1.00**

The middle entries are known as **joint probabilities** and the totals are **marginal probabilities.** In this data set, it appears that more girls than boys take Spanish class. However, that might not be the case because more girls than boys were surveyed and the results might be misleading. To avoid such errors, **conditional relative frequencies** are used. The relative frequencies are calculated based on a row or column.

Here are the conditional relative frequencies using column totals:

Two-Way Frequency Table

	Spanish	French	German	Total
Boys	10	2	8	**20**
Girls	15	12	3	**30**
Total	**25**	**14**	**11**	**50**

Divide each column entry by that column's total

Two-Way Relative Frequency Table (with respect to table total)

	Spanish	French	German	Total
Boys	0.4	0.14	0.73	**0.4**
Girls	0.6	0.86	0.27	**0.6**
Total	**1.00**	**1.00**	**1.00**	**1.00**

Two-way frequency tables can help in making many conclusions about the data. If either the row or column of conditional relative frequencies differs between each row or column of the table, then an association exists between the two categories. For example, in the above tables, the majority of boys are taking German while the majority of girls are taking French. If the frequencies are equal across the rows, there is no association and the variables are labelled as independent. It's important to note that the association does exist in the above scenario, though these results may not occur the next semester when students are surveyed.

When measuring event probabilities, two-way frequency tables can be used to report the raw data and then used to calculate probabilities. If the frequency tables are translated into relative frequency tables, the probabilities presented in the table can be plugged directly into the formulas for conditional probabilities. By plugging in the correct frequencies, the data from the table can be used to determine if events are independent or dependent.

Conditional probability is the probability that event A will happen given that event B has already occurred. An example of this is calculating the probability that a person will eat dessert once they have eaten dinner. This is different than calculating the probability of a person just eating dessert.

The formula for the conditional probability of event A occurring given B is:

$$P(A|B) = \frac{P\ (A \text{ and } B)}{P(B)}$$

It's defined to be the probability of both A and B occurring divided by the probability of event B occurring. If A and B are independent, then the probability of both A and B occurring is equal to $P(A)P(B)$, so $P(A|B)$ reduces to just $P(A)$. This means that A and B have no relationship, and the probability of A occurring is the same as the conditional probability of A occurring given B. Similarly:

$$P(B|A) = \frac{P\ (B \text{ and A })}{P(A)} = P(B)$$

(if A and B are independent)

Two events aren't always independent. For examples, females with glasses and brown hair aren't independent characteristics. There definitely can be overlap because females with brown hair can wear glasses. Also, two events that exist at the same time don't have to have a relationship. For example, even if all females in a given sample are wearing glasses, the characteristics aren't related. In this case, the probability of a brunette wearing glasses is equal to the probability of a female being a brunette multiplied by the probability of a female wearing glasses. This mathematical test of $P(A \cap B) = P(A)P(B)$ verifies that two events are independent.

Conditional probability is the probability that an event occurs given that another event has happened. If the two events are related, the probability that the second event will occur changes if the other event has happened. However, if the two events aren't related and are therefore independent, the first event to occur won't impact the probability of the second event occurring.

Finding Probabilities of Compound Events

The Probability of a Compound Event

A **simple event** consists of only one outcome. The most popular simple event is flipping a coin, which results in either heads or tails. A **compound event** results in more than one outcome and consists of more than one simple event. An example of a compound event is flipping a coin while tossing a die. The result is either heads or tails on the coin and a number from one to six on the die. The probability of a simple event is calculated by dividing the number of possible outcomes by the total number of outcomes. Therefore, the probability of obtaining heads on a coin is $\frac{1}{2}$, and the probability of rolling a 6 on a die is $\frac{1}{6}$. The probability of compound events is calculated using the basic idea of the probability of simple events. If the two events are independent, the probability of one outcome is equal to the product of the probabilities of each simple event. For example, the probability of obtaining heads on a coin and rolling a 6 is equal to $\frac{1}{2} \times \frac{1}{6} = \frac{1}{12}$. The probability of either A or B occurring is equal to the sum of the probabilities minus the probability that both A and B will occur. Therefore, the probability of obtaining either heads on a coin or rolling a 6 on a die is $\frac{1}{2} + \frac{1}{6} - \frac{1}{12} = \frac{7}{12}$. The two events aren't mutually exclusive because they can happen at the same time. If two events are mutually exclusive, and the probability of both events occurring at the same time is zero, the probability of event A or B occurring equals the sum of both probabilities. An example of calculating the probability of two mutually exclusive events is determining the probability of pulling a king or a queen from a deck of cards. The two events cannot occur at the same time.

Representing Sample Spaces for Compound Events

A sample can be broken up into subsets that are smaller parts of the whole. For example, consider a sample population of females. The sample can be divided into smaller subsets based on the characteristics of each female. There can be a group of females with brown hair and a group of females that wear glasses. There also can be a group of females that have brown hair *and* wear glasses. This "and" relates to the **intersection** of the two separate groups of brunettes and those with glasses. Every female in that intersection group has both characteristics.

Similarly, there also can be a group of females that either have brown hair *or* wear glasses. The "or" relates to the union of the two separate groups of brunettes and glasses. Every female in this group has at least one of the characteristics. Finally, the group of females who do not wear glasses can be discussed. This "not" relates to the **complement** of the glass-wearing group. No one in the complement has glasses. **Venn diagrams** are useful in highlighting these ideas. When discussing statistical experiments, this idea can also relate to events instead of characteristics.

Two events aren't always independent. For example, females with glasses and brown hair aren't independent characteristics. There definitely can be overlap because females with brown hair can wear glasses. Also, two events that exist at the same time don't have to have a relationship. For example, even if all females in a given sample are wearing glasses, the characteristics aren't related. In this case, the probability of a brunette wearing glasses is equal to the probability of a female being a brunette multiplied by the probability of a female wearing glasses. This mathematical test of $P(A \cap B) = P(A)P(B)$ verifies that two events are independent.

Designing and Using a Simulation to Generate Frequencies for Compound Events

Simulations are experiments where devices such as coins, cards, and dice are used to generate results representing real outcomes. The simulations are used to estimate probabilities that replace theoretical probabilities in cases where collecting empirical data or using theoretical probability models is improbable or even impossible. Simulation allows experiments to be performed that closely resemble a real scenario. An example of this is determining how likely it is for a family with four children to have all girls. A coin toss can simulate this probability distribution, where heads means having a girl and tails means having a boy. Dice also can be used to simulate this process and calculate a similar experimental probability distribution by assigning even-numbered rolls to girls and odd-numbered rolls to boys. Performing either simulation would be a much less expensive process than surveying a large number of families with four children.

Practice Questions

1. If a car can travel 300 miles in 4 hours, how far can it go in an hour and a half?
   a. 100 miles
   b. 112.5 miles
   c. 135.5 miles
   d. 150 miles

300/4 = 75   75 · 1.5 = 112.5

2. At the store, Jan spends $90 on apples and oranges. Apples cost $1 each and oranges cost $2 each. If Jan buys the same number of apples as oranges, how many oranges did she buy?
   a. 20
   b. 25
   c. 30
   d. 35

1a + 2o = 90
20+40 ≠ 90
30+60 = 90

3. A piggy bank contains 12 dollars' worth of nickels. A nickel weighs 5 grams, and the empty piggy bank weighs 1050 grams. What is the total weight of the full piggy bank? Enter your answer in the box.

1290 grams

4. A construction company is building a new housing development with the property of each house measuring 30 feet wide. If the length of the street is zoned off at 345 feet, how many houses can be built on the street? Enter your answer in the box.

11 houses

5. A train traveling 50 miles per hour takes a trip lasting 3 hours. If a map has a scale of 1 inch per 10 miles, how many inches apart are the train's starting point and ending point on the map?
   a. 14
   b. 12
   c. 13
   d. 15

6. Bernard can make \$80 per day. If he needs to make \$300 and only works full days, how many days will this take? Enter your answer on the number pad.

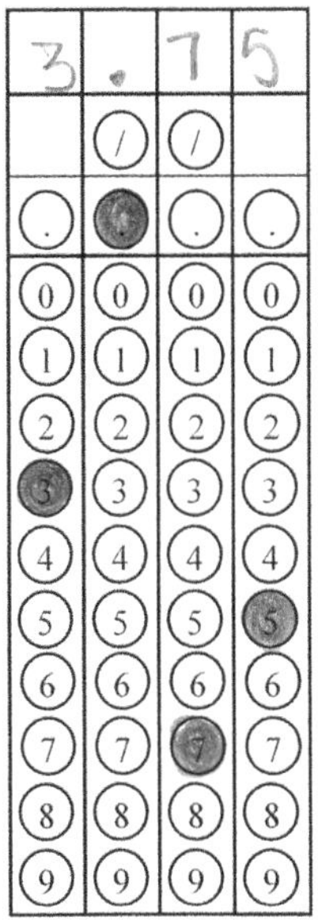

7. What is the probability of randomly picking the winner and runner-up from a race of 4 horses and distinguishing which is the winner?

a. $\frac{1}{4}$

b. $\frac{1}{2}$

c. $\frac{1}{16}$

d. $\frac{1}{12}$

8. A map has a scale of 1 inch per 5 miles. A car can travel 60 miles per hour. If the distance from the start to the destination is 3 inches on the map, how long will it take the car to make the trip? Enter your answer in the box.

15 miles

15 minutes

9. In the following expression, which operation should be completed first? $5 \times 6 + 4 \div 2 - 1$.

a. Multiplication
b. Addition
c. Division
d. Subtraction

10. In order to calculate the perimeter of a legal sized piece of paper that is 14 in. long and $8\frac{1}{2}$ in. wide, what formula would be used?

a. $P = 14 + 8\frac{1}{2}$

(b.) $P = 14 + 8\frac{1}{2} + 14 + 8\frac{1}{2}$

c. $P = 14 \times 8\frac{1}{2}$

d. $P = 14 \times \frac{17}{2}$

11. Using the following diagram, calculate the total circumference, rounding to the nearest decimal place:

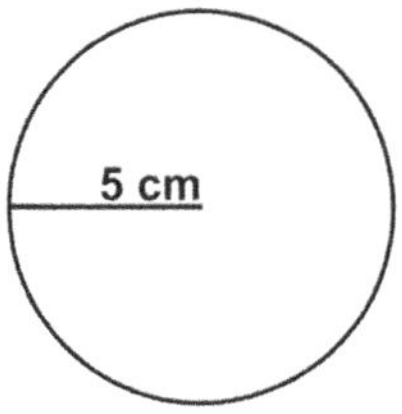

$2\pi r$

$6.28 \cdot 5$

Enter your answer in the box.

31.4cm

12. If Sarah reads at an average rate of 21 pages in four nights, how long will it take her to read 140 pages?

a. 6 nights

b. 26 nights

c. 8 nights

(d.) 27 nights

13. Kassidy drove for 3 hours at a speed of 60 miles per hour. Using the distance formula, $d = r \times t$ ($distance = rate \times time$), how far did Kassidy travel?

a. 20 miles

(b.) 180 miles

c. 65 miles

d. 120 miles

$d = r \times t$

$= 60 \times 3$

$d = 180$ miles

14. The square and circle have the same center. The circle has a radius of r. What is the area of the shaded region?

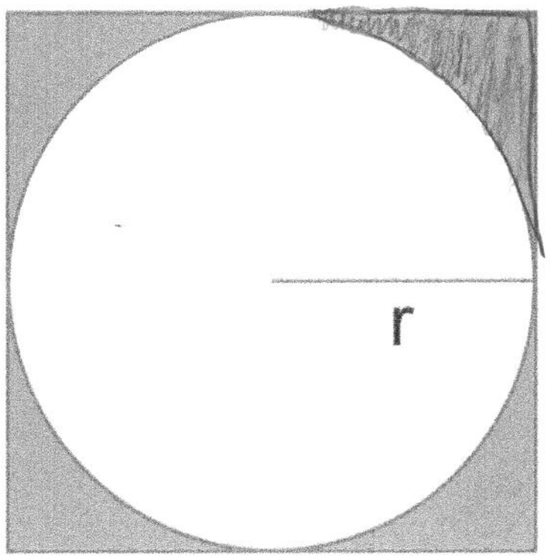

a. $r^2 - \pi r^2$
b. $4r^2 - 2\pi r$
c. $(4 - \pi)r^2$
d. $(\pi - 1)r^2$

15. A truck is carrying three cylindrical barrels. Their bases have a diameter of 2 feet and they have a height of 3 feet. What is the total volume of the three barrels in cubic feet?

a. 3π
b. 9π
c. 12π
d. 15π

16. Greg buys a $10 lunch with 5% sales tax. He leaves a $2 tip after his bill. How much money does he spend?

a. $12.50
b. $12
c. $13
d. $13.25

17. A ball is drawn at random from a ball pit containing 8 red balls, 7 yellow balls, 6 green balls, and 5 purple balls. What's the probability that the ball drawn is yellow?

a. $\frac{1}{26}$
b. $\frac{19}{26}$
c. $\frac{7}{26}$
d. 1

18. Five students take a test. The scores of the first four students are 80, 85, 75, and 60. If the median score is 80, which of the following could NOT be the score of the fifth student?

a. 60
b. 80
c. 85
d. 100

19. An equilateral triangle has a perimeter of 18 feet. If a square whose sides have the same length as one side of the triangle is built, what will be the area of the square?

a. 6 square feet
b. 36 square feet
c. 256 square feet
d. 1000 square feet

20. Nina has a jar where she puts her loose change at the end of each day. There are 13 quarters, 25 dimes, 18 nickels, and 30 pennies in the jar. If she chooses a coin at random, what is the probability that the coin will not be a penny or a dime?

a. 0.36
b. 0.64
c. 0.56
d. 0.34

21. Which fractions are equivalent, or would fill the same portion on a number line? Select all that apply.

a. $\frac{2}{4}$ and $\frac{3}{8}$

b. $\frac{1}{2}$ and $\frac{4}{8}$

c. $\frac{3}{6}$ and $\frac{3}{5}$

d. $\frac{2}{4}$ and $\frac{5}{8}$

e. $\frac{7}{3}$ and $2\frac{1}{3}$

22. Which of the following is a factor of both $x^2 + 4x + 4$ and $x^2 - x - 6$?

a. $x - 3$
b. $x + 2$
c. $x - 2$
d. $x + 3$

23. Which of the following is the correct order of operations?

a. Parentheses, Exponents, Multiplication, Division, Addition, Subtraction
b. Exponents, Parentheses, Multiplication, Division, Addition, Subtraction
c. Parentheses, Exponents, Addition, Multiplication, Division, Subtraction
d. Parentheses, Exponents, Division, Addition, Subtraction, Multiplication

24. If Amanda can eat two times as many mini cupcakes as Marty, what would the missing values be for the following input-output table?

Input (number of cupcakes eaten by Marty)	Output (number of cupcakes eaten by Amanda)
1	2
3	6
5	10
7	14
9	18

a. 6, 10
b. 3, 11
c. 6, 14
d. 4, 12

25. Kelly is selling cookies to raise money for the choir. She has 500 cookies to sell. She sells 45% of the cookies to the sixth graders. At the second lunch, she sells 40% of what's left to the seventh graders. If she sells 60% of the remaining cookies to the eighth graders, how many cookies does Kelly have left at the end of all lunches? Enter your answer on the number pad.

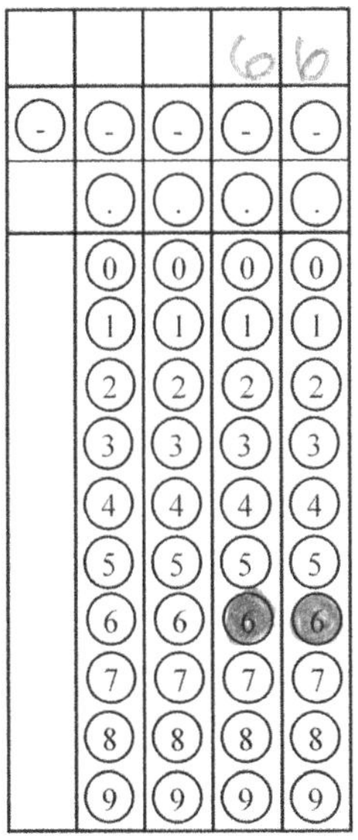

45% 500 = 225
500 - 225 = 275
45% 275
= 110
275 - 110 = 165
= 99
165 - 99
= 66

26. Katie works at a clothing company and sold 192 shirts over the weekend. $\frac{1}{3}$ of the shirts that were sold were patterned, and the rest were solid. Which mathematical expression would calculate the number of solid shirts Katie sold over the weekend?

a. $192 \times \frac{1}{3}$
b. $192 \div \frac{1}{3}$
c. $192 \times (1 - \frac{1}{3})$
d. $192 \div 3$

27. Mo needs to buy enough material to cover the walls around the stage for a theater performance. If he needs 79 feet of wall covering, what is the minimum number of yards of material he should purchase if the material is sold only by whole yards? Enter your answer in the box.

27 yards

28. The soccer team is selling donuts to raise money to buy new uniforms. For every box of donuts they sell, the team receives $3 towards their new uniforms. There are 15 people on the team. How many boxes does each player need to sell in order to raise $270 for their new uniforms? Enter your answer in the box.

6 boxes

29. In the figure below, what is the area of the shaded region?

13 · 4 = 52

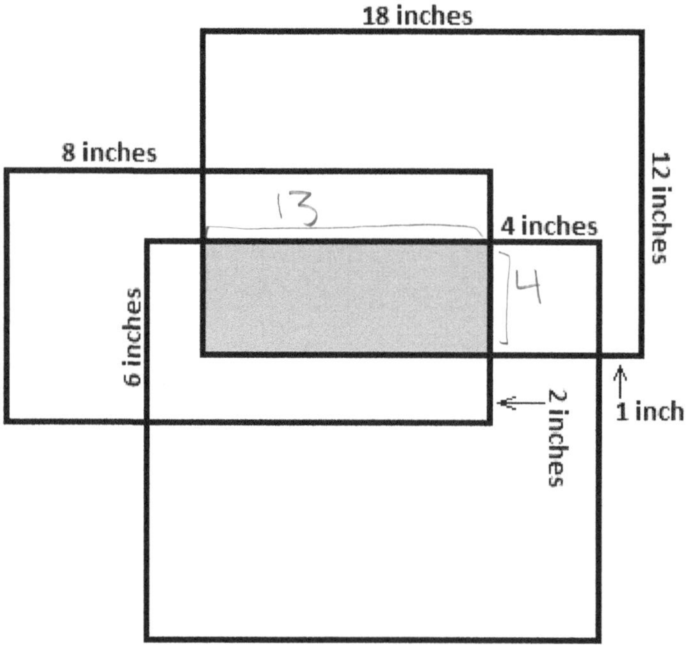

a. 48 sq. inches
b. 52 sq. inches
c. 44 sq. inches
d. 56 sq. inches

30. What is the solution for the following equation?

$$\frac{x^2 + x - 30}{x - 5} = 11$$

a. $x = -6$
b. There is no solution.
c. $x = 16$
d. $x = 5$

31. Which four-sided shape is always a rectangle?

a. Rhombus
b. Square
c. Parallelogram
d. Quadrilateral

32. It costs Shea $12 to produce 3 necklaces. If he can sell each necklace for $20, how much profit would he make if he sold 60 necklaces? Enter your answer in the box.

$960

33. This chart indicates how many sales of CDs, vinyl records, and MP3 downloads occurred over the last year. Approximately what percentage of the total sales was from CDs?

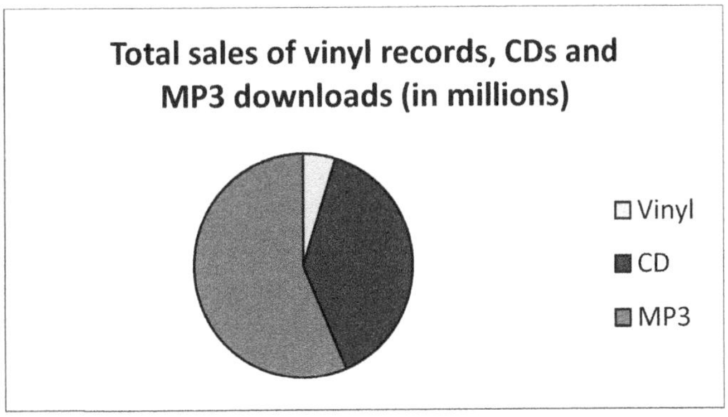

a. 55%
b. 25%
c. 40%
d. 5%

34. If the volume of a sphere is 288π cubic meters, what are the radius and surface area of the same sphere?

a. Radius 6 meters and surface area 144π square meters
b. Radius 36 meters and surface area 144π square meters
c. Radius 6 meters and surface area 12π square meters
d. Radius 36 meters and surface area 12π square meters

35. A closet is filled with red, blue, and green shirts. If there are 54 shirts total, and 12 are green and 28 are red, which equation below could be used to calculate the number of blue shirts?

a. $54 - 12 - 28 = \square$

b. $54 = 28 - 12 + \square$

c. $54 - \square = 28 - 12$

d. $54 + 12 + 28 = \square$

36. Shawna buys $2\frac{1}{2}$ gallons of paint. If she uses $\frac{1}{3}$ of it on the first day, how much does she have left?

a. $1\frac{5}{6}$ gallons

b. $1\frac{1}{2}$ gallons

c. $1\frac{2}{3}$ gallons

d. 2 gallons

$1 - \frac{1}{3} = \frac{2}{3}$ $\quad 2\frac{1}{2} = \frac{5}{2}$

$\frac{2}{3} \times \frac{5}{2} = \frac{10}{6} = 1\frac{4}{6} = 1\frac{2}{3}$

37. The perimeter of a 6-sided polygon is 56 cm. The length of three sides are 9 cm each. The length of two other sides are 8 cm each. What is the length of the missing side?

a. 11 cm

b. 12 cm

c. 13 cm

d. 10 cm

$3(9)=27$, $2(8)=16$; $27 + 16 = 43$; $56 - 43 = 13$

38. The table below shows tickets purchased during the week for entry to the local zoo. What is the mean of adult tickets sold for the week?

Day of the Week	Age	Tickets Sold
Monday	Adult	22
Monday	Child	30
Tuesday	Adult	16
Tuesday	Child	15
Wednesday	Adult	24
Wednesday	Child	23
Thursday	Adult	19
Thursday	Child	26
Friday	Adult	29
Friday	Child	38

a. 24.2

b. 21

c. 22

d. 26.4

$22 + 16 + 24 + 19 + 29 = 110$; $\frac{110}{5} = 22$

39. What unit of volume is used to describe the following 3-dimensional shape?

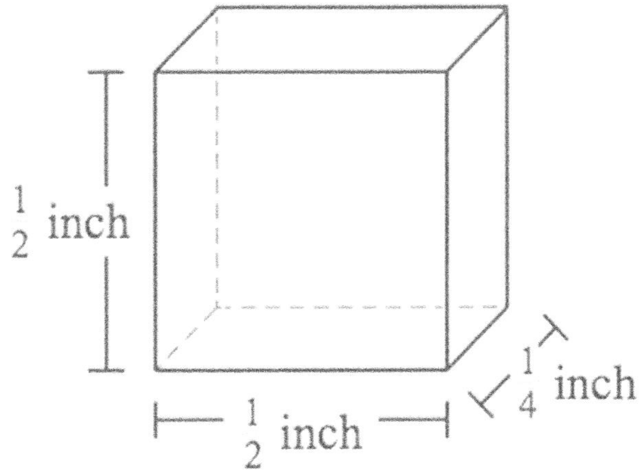

a. Square inches
b. Inches
c. Cubic inches
d. Squares

40. At the beginning of the day, Xavier has 20 apples. At lunch, he meets his sister Emma and gives her half of his apples. After lunch, he stops by his neighbor Jim's house and gives him 6 of his apples. He then uses ¾ of his remaining apples to make an apple pie for dessert at dinner. At the end of the day, how many apples does Xavier have left? Enter your answer in the box.

1 apple

41. There are $4x + 1$ treats in each party favor bag. If a total of $60x + 15$ treats are distributed, how many bags are given out?

a. 15
b. 16
c. 20
d. 22

42. An accounting firm charted its income on the following pie graph. If the total income for the year was $500,000, how much of the income was received from Audit and Taxation Services?

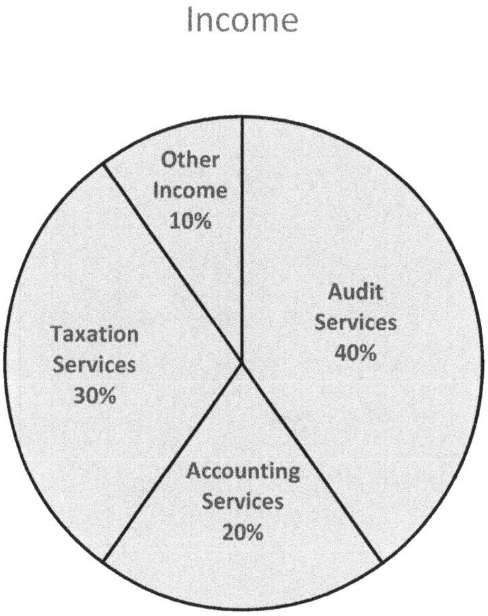

a. $200,000
b. $350,000
c. $150,000
d. $300,000

43. Which inequality represents the number line below?

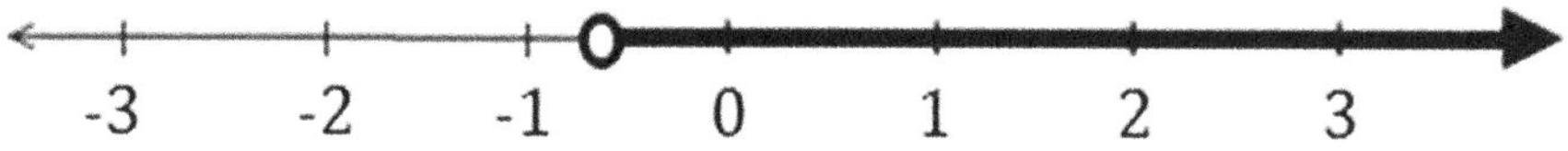

a. $4x + 5 < 8$
b. $-4x + 5 < 8$
c. $-4x + 5 > 8$
d. $4x - 5 > 8$

44. $x^4 - 16$ can be simplified to which of the following?
a. $(x^2 - 4)(x^2 + 4)$

b. $(x^2 + 4)(x^2 + 4)$

c. $(x^2 - 4)(x^2 - 4)$

d. $(x^2 - 2)(x^2 + 4)$

45. Evaluate $9 \times 9 \div 9 + 9 - 9 \div 9$. Enter your answer on the number pad.

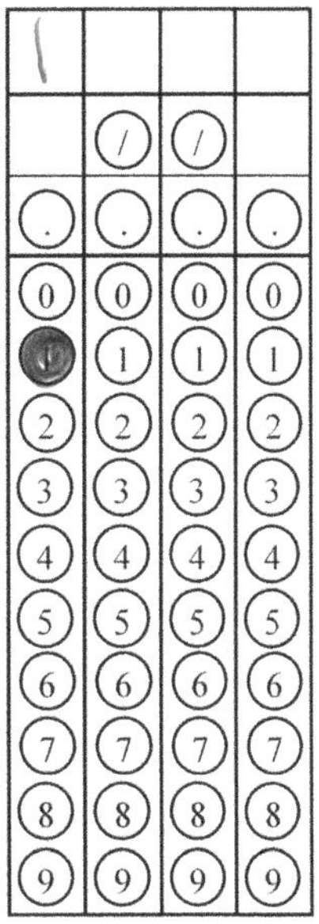

46. If $6t + 4 = 16$, what is t? Enter your answer on the number pad.

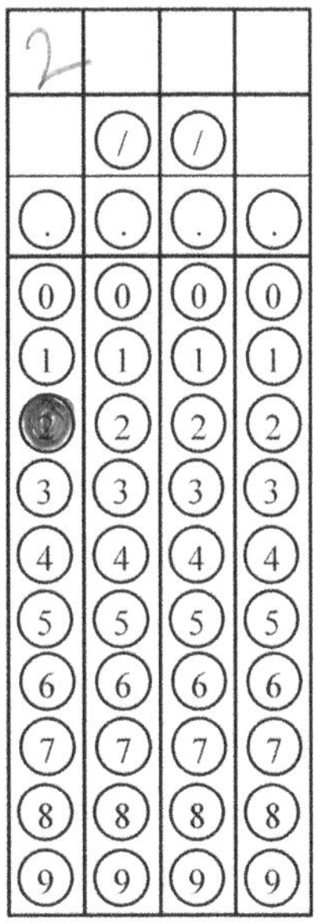

47. Karen gets paid a weekly salary and a commission for every sale that she makes. The table below shows the number of sales and her pay for different weeks.

Sales	2	7	4	8
Pay	\$380	\$580	\$460	\$620

Which of the following equations represents Karen's weekly pay?

a. $y = 90x + 200$
b. $y = 90x - 200$
c. $y = 40x + 300$
d. $y = 40x - 300$

48. The phone bill is calculated each month using the equation $c = 50g + 75$. The cost of the phone bill per month is represented by c, and g represents the gigabytes of data used that month. What is the value and interpretation of the slope of this equation?

a. 75 dollars per day
b. 75 gigabytes per day
c. 50 dollars per day
d. 50 dollars per gigabyte

49. David's car drove 72 miles in 90 minutes. How fast did he drive in feet per second?

a. 0.8 feet per second
b. 48.9 feet per second
c. 0.009 feet per second
d. 70.4 feet per second

50. Jessica buys 10 cans of paint. Red paint costs \$1 per can and blue paint costs \$2 per can. In total, she spends \$16. How many red cans did she buy? Enter your answer in the box.

Answer Explanations

1. B: 300 miles in 4 hours is $300 \div 4 = 75$ miles per hour. In 1.5 hours, the car will go 1.5×75 miles, or 112.5 miles.

2. C: One apple/orange pair costs \$3 total. Therefore, Jan bought $90 \div 3 = 30$ total pairs, and hence, she bought 30 oranges.

3. 2,250 grams: A dollar contains 20 nickels. Therefore, if there are 12 dollars' worth of nickels, there are $12 \times 20 = 240$ nickels. Each nickel weighs 5 grams. Therefore, the weight of the nickels is $240 \times 5 = 1{,}200$ grams. Adding in the weight of the empty piggy bank, the filled bank weighs 2,250 grams.

4. 11: To determine the number of houses that can fit on the street, the length of the street is divided by the width of each house: $345 \div 30 = 11.5$. Although the mathematical calculation of 11.5 is correct, this answer is not reasonable. Half of a house cannot be built, so the company will need to either build 11 or 12 houses. Since the width of 12 houses (360 feet) will extend past the length of the street, only 11 houses can be built.

5. D: First, the train's journey in the real word is $3 \times 50 = 150$ miles. On the map, 1 inch corresponds to 10 miles, so there is $150 \div 10 = 15$ inches on the map.

6.

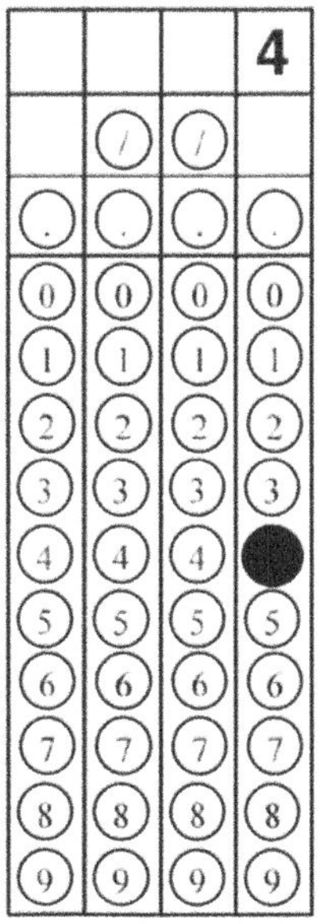

300/80 =30/8 = 15/4 =3.75. But Bernard is only working full days, so he will need to work 4 days, since 3 days are not sufficient.

7. D: $\frac{1}{12}$. The probability of picking the winner of the race is $\frac{1}{4}\left(\frac{number\ of\ favorable\ outcomes}{number\ of\ total\ outcomes}\right)$. Assuming the winner was picked on the first selection, three horses remain from which to choose the runner-up (these are dependent events). Therefore, the probability of picking the runner-up is $\frac{1}{3}$. To determine the probability of multiple events, the probability of each event is multiplied: $\frac{1}{4} \times \frac{1}{3} = \frac{1}{12}$.

8. 15 minutes: The journey will be $5 \times 3 = 15$ miles. A car traveling at 60 miles per hour is traveling at 1 mile per minute. So, it will take $15 \div 1 = 15$ minutes to take the journey.

second digit to the right of the decimal point is equal to the value $\div$ (10×10), or the value $\div$ 100.

9. A: Using the order of operations, multiplication and division are computed first from left to right. Multiplication is on the left; therefore, the multiplication should be performed first.

10. B: Perimeter of a rectangle is the sum of all four sides. Therefore, the answer is:

$$P = 14 + 8\frac{1}{2} + 14 + 8\frac{1}{2}$$

$$14 + 14 + 8 + \frac{1}{2} + 8 + \frac{1}{2} = 45 \text{ square inches}$$

11. 31.4 sq. cm: To calculate the circumference of a circle, use the formula $2\pi r$, where r equals the radius or half of the diameter of the circle and $\pi = 3.14 \dots$. Substitute the given information, $2\pi 5 = 31.4$.

12. D: This problem can be solved by setting up a proportion involving the given information and the unknown value. The proportion is:

$$\frac{21\ pages}{4\ nights} = \frac{140\ pages}{x\ nights}$$

Solving the proportion by cross-multiplying, the equation becomes $21x = 4 \times 140$, where $x = 26.67$. Since it is not an exact number of nights, the answer is rounded up to 27 nights. Twenty-six nights would not give Sarah enough time.

13. B: 180 miles. The rate, 60 miles per hour, and time, 3 hours, are given for the scenario. To determine the distance traveled, the given values for the rate (*r*) and time (*t*) are substituted into the distance formula and evaluated:

$$d = r \times t$$

$$d = (60mi/h) \times (3h) \rightarrow d = 180mi$$

14. C: The area of the shaded region is the area of the square, minus the area of the circle. The area of the circle will be πr^2. The side of the square will be $2r$, so the area of the square will be $4r^2$. Therefore, the difference is:

$$4r^2 - \pi r^2 = (4 - \pi)r^2$$

15. B: The formula for the volume of a cylinder is $\pi r^2 h$, where *r* is the radius and *h* is the height. The diameter is twice the radius, so these barrels have a radius of 1 foot. That means each barrel has a volume of:

$$\pi \times 1^2 \times 3 = 3\pi \text{ cubic feet}$$

Since there are three of them, the total is $3 \times 3\pi = 9\pi$ cubic feet.

16. A: The tip is not taxed, so he pays 5% tax only on the $\$10$. 5% of $\$10$ is $0.05 \times 10 = \$0.50$. Add up $\$10 + \$2 + \$0.50$ to get $\$12.50$.

17. C: The sample space is made up of $8 + 7 + 6 + 5 = 26$ balls. The probability of pulling each individual ball is $\frac{1}{26}$. Since there are 7 yellow balls, the probability of pulling a yellow ball is $\frac{7}{26}$.

18. A: Lining up the given scores provides the following list: 60, 75, 80, 85, and one unknown. Because the median needs to be 80, it means 80 must be the middle data point out of these five. Therefore, the unknown data point must be the fourth or fifth data point, meaning it must be greater than or equal to 80. The only answer that fails to meet this condition is 60.

19. B: An equilateral triangle has three sides of equal length, so if the total perimeter is 18 feet, each side must be 6 feet long. A square with sides of 6 feet will have an area of $6^2 = 36$ square feet.

20. A: The total number of coins in the jar is 86, which is the sum of all the coins. The probability of Nina choosing a coin other than a penny or a dime can be found by calculating the total of quarters and nickels. This total is 31. Taking 31 and dividing it by 86 gives the probability of choosing a coin that is not a penny or a dime. This decimal found from the fraction $\frac{31}{86}$ is 0.36.

21. B & E: $\frac{1}{2}$ is the same fraction as $\frac{4}{8}$, and would both fill up the same portion of a number line.

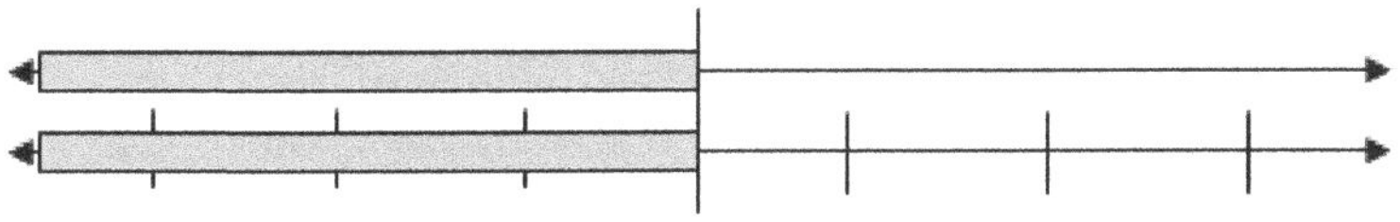

The other answer choice that contains equivalent fractions is *E*, $\frac{7}{3}$ and $2\frac{1}{3}$. $\frac{7}{3}$ is the improper fraction for the mixed number $2\frac{1}{3}$. Recall that to convert a mixed number to an improper fraction, multiply the denominator by the whole number then add the numerator and place this sum over the denominator (as your new numerator). In this case:

$$2\frac{1}{3} \rightarrow 3 \times 2 + 1 \rightarrow \frac{7}{3}$$

22. B: To factor $x^2 + 4x + 4$, the numbers needed are those that add to 4 and multiply to 4. Therefore, both numbers must be 2, and the expression factors to:

$$x^2 + 4x + 4 = (x + 2)^2$$

Similarly, the second expression factors to $x^2 - x - 6 = (x - 3)(x + 2)$, so that they have $x + 2$ in common.

23. A: Order of operations follows PEMDAS—Parentheses, Exponents, Multiplication and Division from left to right, and Addition and Subtraction from left to right.

24. C: The situation can be described by the equation $? \times 2$. Filling in for the missing numbers would result in $3 \times 2 = 6$ and $7 \times 2 = 14$. Therefore, the missing numbers are 6 and 14. The other choices are miscalculations or misidentification of the pattern formed by the table.

25.

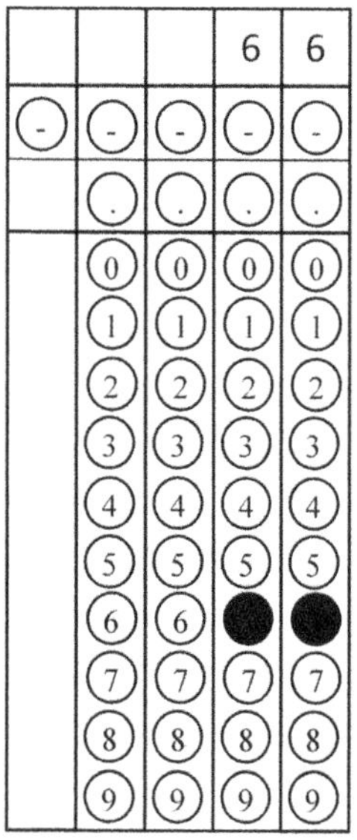

66 cookies. If the sixth graders bought 45% of the cookies, the number they bought is found by multiplying 0.45 by 500. They bought 225 cookies. The number of cookies left is $500 - 225 = 275$. During the second lunch, the seventh graders bought 40% of the cookies, which is found by multiplying 0.40 by the remaining 275 cookies. The seventh graders bought 110 cookies. This leaves 165 cookies to sell to the eighth graders. If they bought 60% of the remaining cookies, then they bought 99 cookies. Subtracting 99 from 165 cookies leaves Kelly with 66 cookies remaining after the three lunches.

26. C: $\frac{1}{3}$ of the shirts sold were patterned. Therefore, $1 - \frac{1}{3} = \frac{2}{3}$ of the shirts sold were solid. Anytime "of" a quantity appears in a word problem, multiplication needs to be used. Therefore:

$$192 \times \frac{2}{3} = 192 \times \frac{2}{3} = \frac{384}{3} = 128 \text{ solid shirts were sold}$$

The entire expression is $192 \times \left(1 - \frac{1}{3}\right)$.

27. 27 yards: In order to solve this problem, the number of feet in a yard must be established. There are 3 feet in every yard. The equation to calculate the minimum number of yards is:

$$79 \div 3 = 26\frac{1}{3}$$

If the material is sold only by whole yards, then Mo would need to round up to the next whole yard in order to cover the extra $\frac{1}{3}$ yard. Therefore, the answer is 27 yards.

28. 6 boxes: The team needs a total of \$270, and each box earns them \$3. Therefore, the total number of boxes needed to be sold is $270 \div 3$, which is 90. With 15 people on the team, the total of 90 can be divided by 15, which equals 6. This means that each member of the team needs to sell 6 boxes for the team to raise enough money to buy new uniforms.

29. B: This can be determined by finding the length and width of the shaded region. The length can be found using the length of the top rectangle which is 18 inches, then subtracting the extra length of 4 inches and 1 inch. This means the length of the shaded region is 13 inches. Next, the width can be determined using the 6 inch measurement and subtracting the 2 inch measurement. This means that the width is 4 inches. Thus, the area is $13 \times 4 = 52$ sq. inches.

30. B: The equation can be solved by factoring the numerator into $(x + 6)(x - 5)$. Since that same factor $(x - 5)$ exists on top and bottom, that factor cancels. This leaves the equation $x + 6 = 11$. Solving the equation gives the answer $x = 5$. When this value is plugged into the equation, it yields a zero in the denominator of the fraction. Since this is undefined, there is no solution.

31. B: A rectangle is a specific type of parallelogram. It has 4 right angles. A square is a rhombus that has 4 right angles. Therefore, a square is always a rectangle because it has two sets of parallel lines and 4 right angles.

32. \$960: In order to calculate the profit, an equation modeling the total income less the cost of the materials needs to be formulated. $\$60 \times 20 = \$1{,}200$ total income. $60 \div 3 = 20$ sets of materials. $20 \times \$12 = \$240$ cost of materials.

$$\$1{,}200 - \$240 = \$960 \text{ profit}$$

33. C: The sum total percentage of a pie chart must equal 100%. Since the CD sales take up less than half of the chart and more than a quarter (25%), it can be determined to be 40% overall. This can also be measured with a protractor. The angle of a circle is 360°. Since 25% of 360 would be 90° and 50% would be 180°, the angle percentage of CD sales falls in between; therefore, it would be answer *C*.

34. A: Because the volume of the given sphere is 288π cubic meters, this means ${}^{4}/{}_{3}\,\pi r^3 = 288\pi$. This equation is solved for r to obtain a radius of 6 meters. The formula for the surface area of a sphere is $4\pi r^2$, so if $r = 6$ in this formula, the surface area is 144π square meters.

35. A: The sum of the shirts is 54. So, we know that the number of green shirts, plus the number of red shirts, plus the number of blue shirts is 54. By subtracting the number of green shirts (12) and the number of red shirts (28) from the total number of shirts (54), we can calculate the number of blue shirts, which make up the remaining portion of the total.

36. C: If she has used 1/3 of the paint, she has 2/3 remaining. $2\frac{1}{2}$ gallons are the same as $\frac{5}{2}$ gallons. The calculation is:

$$\frac{2}{3} \times \frac{5}{2} = \frac{5}{3} = 1\frac{2}{3} \text{ gallons}$$

37. C: The perimeter is found by calculating the sum of all sides of the polygon.

$$9 + 9 + 9 + 8 + 8 + s = 56, \text{ where } s \text{ is the missing side length}$$

Therefore, 43 plus the missing side length is equal to 56. The missing side length is 13 cm.

Perimeter is found by calculating the sum of all sides of the polygon. $9+9+9+8+8+s=56$, where *s* is the missing side length. Therefore, 43 plus the missing side length is equal to 56. The missing side length is 13 cm.

38. C: To find the mean, or average, of a set of values, add the values together and then divide by the total number of values. Each day of the week has an adult ticket amount sold that must be added together. The equation is as follows:

$$\frac{22+16+24+19+29}{5}=22$$

39. C: Volume of this three-dimensional figure is calculated using length x width x height. Each measure of length is in inches. Therefore, the answer would be labeled in cubic inches.

40. 1: This problem can be solved using basic arithmetic. Xavier starts with 20 apples, then gives his sister half, so 20 divided by 2.

$$\frac{20}{2}=10$$

He then gives his neighbor 6, so 6 is subtracted from 10.

$$10-6=4$$

Lastly, he uses ¾ of his apples to make an apple pie, so to find remaining apples, the first step is to subtract ¾ from one and then multiply the difference by 4.

$$\left(1-\frac{3}{4}\right)\times 4=?$$

$$\left(\frac{4}{4}-\frac{3}{4}\right)\times 4=?$$

$$\left(\frac{1}{4}\right)\times 4=1$$

41. A: Each bag contributes $4x+1$ treats. The total treats will be in the form $4nx+n$ where *n* is the total number of bags. The total is in the form $60x+15$, from which it is known $n=15$.

42. B: $350,000: Since the total income is $500,000, then a percentage of that can be found by multiplying the percent of Audit Services as a decimal, or 0.40, by the total of 500,000. This answer is found from the equation:

$$500000\times 0.4=200000$$

The total income from Audit Services is $200,000.

For the income received from Taxation Services, the following equation can be used:

$$500000\times 0.3=150000$$

The total income from Audit Services and Taxation Services is:

$$150{,}000 + 200{,}000 = 350{,}000$$

Another way of approaching the problem is to calculate the easy percentage of 10%, then multiply it by 7, because the total percentage for Audit and Taxation Services was 70%. 10% of 500,000 is 50,000. Then multiplying this number by 7 yields the same income of $350,000.

43. B: The number line shows:

$$x > -\frac{3}{4}$$

Each inequality must be solved for x to determine if it matches the number line. Choice *A* of $4x + 5 < 8$ results in $x < -\frac{3}{4}$, which is incorrect. Choice *C* of $-4x + 5 > 8$ yields $x < -\frac{3}{4}$, which is also incorrect. Choice *D* of $4x - 5 > 8$ results in $x > \frac{13}{4}$, which is not correct. Choice B, $-4x + 5 < 8$ is the only choice that results in the correct answer of:

$$x > -\frac{3}{4}$$

44. A: This has the form $t^2 - y^2$, with $t = x^2$ and $y = 4$. It's also known that $t^2 - y^2 = (t + y)(t - y)$, and substituting the values for *t* and *y* into the right-hand side gives:

$$(x^2 - 4)(x^2 + 4)$$

45.

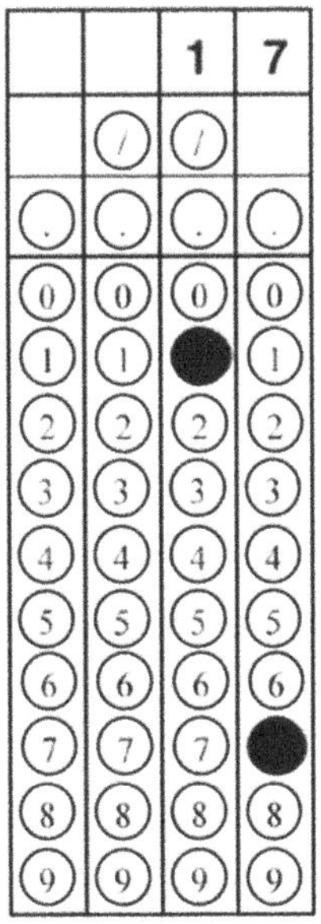

According to order of operations, multiplication and division must be completed first from left to right. Then, addition and subtraction is completed from left to right. Therefore:

$$9 \times 9 \div 9 + 9 - 9 \div 9$$

$$81 \div 9 + 9 - 9 \div 9$$

$$9 + 9 - 1$$

$$18 - 1$$

$$17$$

46.

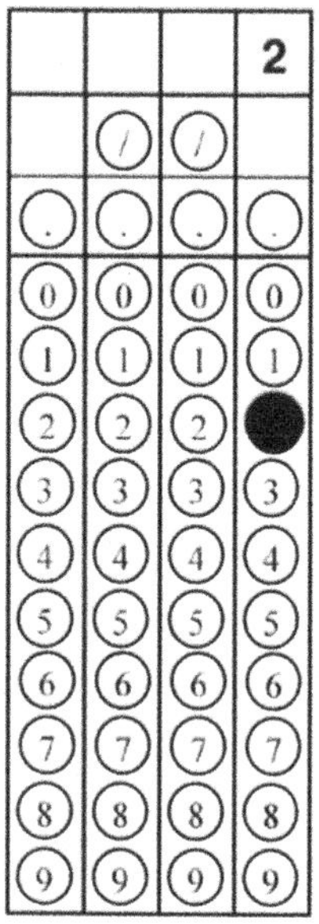

First, subtract 4 from each side. This yields $6t = 12$. Now, divide both sides by 6 to obtain $t = 2$.

47. C: $y = 40x + 300$. In this scenario, the variables are the number of sales and Karen's weekly pay. The weekly pay depends on the number of sales. Therefore, weekly pay is the dependent variable (y), and the number of sales is the independent variable (*x*). Each pair of values from the table can be written as an ordered pair (*x*, *y*): (2,380), (7,580), (4,460), (8,620). The ordered pairs can be substituted into the equations to see which creates true statements (both sides equal) for each pair. Even if one ordered pair produces equal values for a given equation, the other three ordered pairs must be checked.

The only equation which is true for all four ordered pairs is $y = 40x + 300$:

$$380 = 40(2) + 300 \rightarrow 380 = 380$$

$$580 = 40(7) + 300 \rightarrow 580 = 580$$

$$460 = 40(4) + 300 \rightarrow 460 = 460$$

$$620 = 40(8) + 300 \rightarrow 620 = 620$$

48. D: The slope from this equation is 50, and it is interpreted as the cost per gigabyte used. Since the g-value represents number of gigabytes and the equation is set equal to the cost in dollars, the slope relates these two values. For every gigabyte used on the phone, the bill goes up 50 dollars.

49. D: This problem can be solved by using unit conversions. The initial units are miles per minute. The final units need to be feet per second. Converting miles to feet uses the equivalence statement 1 mile =

5,280 feet. Converting minutes to seconds uses the equivalence statement 1 minute = 60 seconds. Setting up the ratios to convert the units is shown in the following equation:

$$\frac{72\ miles}{90\ minutes} * \frac{1\ minute}{60\ seconds} * \frac{5280\ feet}{1\ mile} = 70.4 \text{ feet per second}$$

The initial units cancel out, and the new, desired units are left.

50. 4: We are trying to find x, the number of red cans. The equation can be set up like this:

$$x + 2(10 - x) = 16$$

The left x is actually multiplied by \$1, the price per red can. Since we know Jessica bought 10 total cans, $10 - x$ is the number blue cans that she bought. We multiply the number of blue cans by \$2, the price per blue can.

That should all equal \$16, the total amount of money that Jessica spent. Working that out gives us:

$$x + 20 - 2x = 16$$

$$20 - x = 16$$

$$x = 4$$

Dear Customer,

Thank you for purchasing this book. We hope that we exceeded your expectations.

Our goal in creating this book was to cover all of the topics that are likely to appear on the test. We also strove to make our practice questions as similar as possible to what will be seen on test day. With that being said, if you found something that you feel was not up to your standards, please send us an email and let us know.

We have study guides in a wide variety of fields. If you're interested in one, try searching for it on Amazon or send us an email.

Thanks Again and Happy Testing!
Product Development Team
info@studyguideteam.com

Interested in buying more than 10 copies of our product? Contact us about bulk discounts:

bulkorders@studyguideteam.com

FREE Test Taking Tips DVD Offer

To help us better serve you, we have developed a Test Taking Tips DVD that we would like to give you for FREE. **This DVD covers world-class test taking tips that you can use to be even more successful when you are taking your test.**

All that we ask is that you email us your feedback about your study guide. Please let us know what you thought about it – whether that is good, bad or indifferent.

To get your **FREE Test Taking Tips DVD**, email freedvd@studyguideteam.com with "FREE DVD" in the subject line and the following information in the body of the email:

a. The title of your study guide.

b. Your product rating on a scale of 1-5, with 5 being the highest rating.

c. Your feedback about the study guide. What did you think of it?

d. Your full name and shipping address to send your free DVD.

If you have any questions or concerns, please don't hesitate to contact us at freedvd@studyguideteam.com.

Thanks again!

Made in the USA
Monee, IL
28 June 2020

35248340R00063